Beta-Glucan: Sources, Properties and Applications

Mudasir Bashir Mir • Shabir Ahmad Mir
Editors

# Beta-Glucan: Sources, Properties and Applications

Springer

*Editors*
Mudasir Bashir Mir
Department of Food Science and Technology
Govt. Degree College
Baramulla, Jammu and Kashmir, India

Shabir Ahmad Mir
Department of Food Science and Technology
Govt. College for Women
M. A. Road, Srinagar
Jammu and Kashmir, India

ISBN 978-3-031-95787-1 ISBN 978-3-031-95788-8 (eBook)
https://doi.org/10.1007/978-3-031-95788-8



This Springer imprint is published by the registered company Springer Nature Switzerland AG
The registered company address is: Gewerbestrasse 11, 6330 Cham, Switzerland

# Preface

Beta-glucans, a class of non-starch polysaccharides, have gained popularity due to their remarkable biological and functional properties. Derived from diverse sources such as oats, barley, mushrooms, and yeast, β-glucan has gained significant attention for its potential to transform the landscape of foods, medicines, cosmetics, and health products. The source and structure of the β-glucans have the leading role in affecting their functional and nutritional quality. Beta-glucan, being a bioactive polysaccharide, has emerged as a potential solution for preventing or treating disorders, such as reduces cholesterol and triglyceride levels, maintains blood glucose homeostasis, and improves immunity. It is also known for its anti-cancerous and antioxidant properties.

This book, *Beta-Glucan: Sources, Properties, and Applications*, is intended to be a comprehensive reference book covering fundamental concepts and the latest developments. It is designed to serve as a valuable resource for researchers, industry professionals, and undergraduate and postgraduate students of food science and technology, pharmaceutical science, food engineering, and other associated fields. The book is first of its kind in providing details on the sources, main methods of extraction, chemistry, health benefits of beta-glucans, stability during processing, industrial applications, and bioactive delivery systems. We hope that this comprehensive book serves as an informative guide and inspires further advancements in the study and utilization of beta-glucans. We would like to acknowledge the sincere contributions from all the authors whose expertise has made this work possible. The assistance from Springer Nature is gratefully appreciated in publishing the book.

Baramulla, Jammu and Kashmir, India — Mudasir Bashir Mir
Srinagar, Jammu and Kashmir, India — Shabir Ahmad Mir

# Contents

# About the Editors

**Mudasir Bashir Mir, PhD** is currently working as an Assistant Professor at the Government Degree College Baramulla. He holds a Master of Science (MSc) degree in Food Technology from the University of Kashmir, India, and PhD in Food Science and Technology from Pondicherry University, Puducherry, India. He has contributed significantly to Food Science through his extensive research on the modification of food polysaccharides and their applications in food and nutraceutical industries. He has published numerous research articles and book chapters in reputed scientific journals. He has collaborated with national and international scholars in advancing Food technology and continues to contribute to the field through his teaching and research.

**Shabir Ahmad Mir** is an Assistant Professor at the Government College for Women, M. A. Road, Srinagar, Jammu and Kashmir, India. He obtained his MSc (Food Technology) degree from Islamic University of Science and Technology, Awantipora, India, and PhD (Food Technology) from Pondicherry University, Puducherry, India. He received the best PhD Thesis Award for outstanding research in the field of food technology. He is also editorial board member of several journals. He has organized several conferences and workshops in Food Science and Technology. He has published numerous international papers, book chapters, and edited 15 books.

# Chapter 1
# Introduction to Beta-Glucan

**Mudasir Bashir Mir, Shabir Ahmad Mir, and Sameer Ahmad**

## 1.1 Introduction

β-Glucan is a soluble and non-digestible homopolysaccharide. It is a dietary fiber primarily found in cereals, barley, seaweed, and various bacteria as well as fungi such as *Aspergillus fumigatus, Candida albicans,* and *Saccharomyces cerevisiae* (Ye et al., 2022). β-glucan accumulates in the inner portion of yeast's cell wall and the aleurone portion of cereals, where it aids in forming the cell wall. β-Glucan comprises over 250,000 D-glucopyranose units linked through glycosidic bonds (Vannucci et al., 2013). β-(1 → 3)-bond is the backbone of the structure with side chains attached that could be either β-(1 → 4) in the case of cereals or β-(1 → 6) in the case of fungi) Han et al. (2020). There is also another complex structure containing cyclic β-(1 → 3, 1 → 6) and hyperbranched β-(1 → 3, 1 → 4, 1 → 6) present in a bacteria belonging to the *Rhizobiaceae* family and *Pleurotus tuber-region* mushroom sclerotia (Breedveld & Miller, 1994). Depending on the source, the structure varies according to the molecular structure, solubility, type of linkage, and chain conformation. A study on wild barley and farmed barley (Nishantha et al., 2018) found the amounts of β-glucan to be 3.27–6.67% and 2.68–4.74%, respectively. It was discovered that waxy-type barley or barley with the hull present had greater levels of β-glucan than hulled and non-waxy barley (Islam et al., 2021). By merging Glacier AC 38 with the Shikoku hadaka type of barley, Korea's Rural Development

M. B. Mir (✉)
Department of Food Science & Technology, Govt. Degree College, Baramulla, Jammu & Kashmir, India

S. A. Mir
Department of Food Science & Technology, Govt. College for Women, M. A. Road, Srinagar, Jammu & Kashmir, India

S. Ahmad
Department of Bioengineering, Integral University, Lucknow, India



M. Bashir Mir, S. A. Mir (eds.), *Beta-Glucan: Sources, Properties and Applications*,
https://doi.org/10.1007/978-3-031-95788-8_1

Administration (RDA) has created a new variety of barley known as Betaone, which has a greater concentration of β-glucan than previous kinds.

The presence of multiple hydroxyl clusters in the β-glucan fosters the linking with reactive groups of other molecules and alters the solubility, conformation, and aggregation ability. The MW of the beta-glucan is influenced by the position, degree, and length of branching and the branches' structure, shape, and size. These inherent features are vital in making the beta-glucan complete diverse functional roles in food products and contribute to the biological functions in the living system (Kaur et al., 2019). The physical and physicochemical characteristics of beta-glucans hold considerable importance in entrepreneurship and nutrition. Based on their flow characteristics, beta-glucans have also been considered for use as food hydrocolloids. Under specific circumstances, β-glucans have demonstrated the gel-forming potential and increased viscosity in liquids. Beta-glucans can improve the viscosity of various food formulations by acting as thickeners; alternatively, they can be employed as fat substitutes to produce low-calorie foods. In addition, because of their advantageous physiological effects, β-glucans from cereals have recently been the subject of numerous attempts to increase their content in food formulations. So far, products like breakfast cereals, spaghetti, noodle dishes, baked items, dairy, and meat products have been successfully incorporated with ample fractions of beta-glucan from cereals. The linking arrangements, conformity, and molecular weight or distributions of beta-glucans generally govern their physical properties, including their solubility and rheological conduct in suspension and gel conditions (Lazaridou et al., 2007).

## 1.2 Sources

There are many sources for the production of beta-glucan (BG), such as cereals, yeast, bacteria, and mold. The source and structure of the β-glucan have the leading role in affecting their functional and nutritional quality. The literature reported that the total β-glucan content is higher in cereals than in microbial sources (Choi & Kim, 2023). The digestibility of the cereal BG was reported to be in the range of 63–73%. However, the total digestibility was found in the range of 89–93% (Choi & Kim, 2023). Similarly, the β-glucan obtained from the microbial sources (*Saccharomyces cerevisiae*) shows a 53–83% insoluble fraction, and the main reason behind the difference in the range is the structural difference of bonds in the β-glucan (Suchecka et al., 2017). However, in the other non-cereal samples like yeast, bacteria, algae, and mushrooms, β-Glucan is present inside the sample matrix as a fibrous structure (Ahmad & Khalid, 2018). Apart from the fibrous structures, these compounds are linked in the linkage of 1–3 β-glycosidic and 1–6 β-glycosidic linkages, which differs from the linkage in cereal grains. Glucans are the main components with the major share in constituting the fungi and yeast cell walls (Cerletti et al., 2021). The difference in the linkage of the cereal and non-cereal samples is the main factor that affects the quality of the BG. Similarly, the different properties observed for BG in terms of chemical, molecular, and functional characteristics like solubility digestibility were influenced greatly (Zhu et al., 2016). On the other hand, factors like primary structure, polymer charge, solubility, degree of branching, and molecular weight play important role in providing the biological functional values of the β-glucan (Muthuramalingam et al., 2021; Shoukat & Sorrentino, 2021).

## 1.3 Extraction Methods

Substantial research reported so far on techniques for both qualitative and quantitative extraction and the ensuing isolation of beta-glucan from natural sources, a comprehensive grasp of the selection of appropriate techniques and their alterations for further improvements in both quality and quantity can be achieved (Ahmad et al., 2009; Bhatty, 1995; Carr et al., 1990; Irakli et al., 2004; Papageorgiou et al., 2005; Skendi et al., 2003). The methods used to extract beta-glucan from these sources can be broadly classified into three major classes: enzymatic, physical, and chemical. The chemical extraction process entails dissolving the source material's cell walls and releasing the beta-glucan using alkali, acids, and solvents. In the acid extraction strategy (Ahluwalia & Ellis, 1984; Babu, 2015; Maheshwari et al., 2017), the source material is treated with an acid solution (citric acid) to extract the beta-glucan or an alkaline solution (NaOH) in the alkaline extraction method. As described by Irakli et al. (2004) and Papageorgiou et al. (2005), the hot water extraction approach entails heating the source material to different temperatures and times to extract the beta-glucan. Proteases and amylases are specific enzymes used in the enzymatic extraction process to eliminate protein and starch impurities from the desired product. This method is commonly used because of its selectivity and ability to produce high-quality beta-glucan with minimal damage to the source material (Babu, 2015; Carr et al., 1990; Izydorczyk et al., 2000). Physical extraction methods that use physical forces to break down cell walls and extract beta-glucan include high-pressure homogenization, ultrasonication, and microfluidization (Izydorczyk et al., 2000; Tian et al., 2012; Zhu et al., 2016).

## 1.4 Chemistry

A non-starch polysaccharide, beta-glucan is made up of linear D-glucose chains with a (1–3)-β-D-glycosidic linkage and side chains that are connected to β-(1–6) in fungi and β-(1–4) in plants and bacteria. Oats include a random unbranched linear structure of β-(1–3) and β-(1–4)-D-glycosidic linkage. It contains 70% of β-(1–4) and 30% of β-(1–3)-glycosidic bonds in its structure. The presence of β-(1–3) linkage breaks the uniform structure of oat beta-glucan, allowing it to form a viscous solution upon solubilization (Xu et al., 2013). The structure of wheat beta-glucan is quite regular due to the predominance of trimeric unit that contributes to its high gelling ability. Beta-glucan derived from yeast and mushroom composed of β-(1–3) linkage with occasional β-(1–6) linkages has greater biological activity than cereal beta-glucan (β-(1–3) and β-(1–4)). Fungal beta-glucan is a mixture of linear β-(1–3) linkage with short branch chains connected through β-(1–6) (Nakashima et al., 2018), while bacterial beta-glucan has a linear unbranched chain of β-(1–3)-D-glucan backbone. Seaweed/algae beta-glucan comprises a straight chain of β-(1–3)-D-glucan with high levels of β-(1–6) branches. Beta-glucan is divided into two classes, soluble having β-(1–3, 1–6)-D-linked glucose and insoluble having β-(1–3, 1–4)-D-linked glucose. Based on the source, beta-glucan can exist in various conformations such as random coils, helices (single, double, or triple), rod-like, and wormlike shapes.

## 1.5 Physicochemical Properties

The physicochemical attributes of beta-glucan can be affected by isolation processes, including temperature, pH, particle size, and solvents. β-glucan exhibits an excellent role in food processing due to its emulsifying, thickening, and water-holding capacities (Liu et al., 2021). Beta-glucan's physicochemical features are crucial in commercial and dietary uses. Beta-glucan utilization as food hydrocolloids has been investigated by understanding their rheological traits. They can form a gel in specific conditions and improve the product's viscosity. The ability of the beta-glucan to change the texture and expression of food commodities makes them appropriate as thickening agents. The beta-glucan's molar ratio and molar mass varied with sources and significantly contributed to the rheological behavior. Molecular weight, concentration, temperature, pH, and agronomic conditions influence beta-glucan's physicochemical and functional attributes. The physical characteristics of β-glucan, including solubility and rheological conduct in suspensions and gel phases, are modified based on linking patterns, structure, and molecular weight or distributions. These characteristics may be modified using various techniques and variables, opening up new possibilities for their use in various food, pharmaceutical, and cosmetological applications. However, to better manage and preserve the chemical-physical properties of β-glucans, it is necessary to pay close attention to the processes of extraction, processing, manipulation, and preservation of food, which subject this functional polymer to a series of stress factors. These conditions threaten its chemical and functional characteristics (Lante et al., 2023).

## 1.6 Health Benefits

Recent research has also shown that changes in the gut microbiota environment are linked to a positive reduction in Metabolic Syndrome (MS) (Chanmuang et al., 2022; Chen et al., 2022b; Shehzad et al., 2023). Beta-glucan (BG) has emerged as a potential solution for treating these disorders, as it reduces cholesterol and triglyceride levels, maintains blood glucose homeostasis, and improves immunity. Beta-glucan is also known for its anti-cancerous and antioxidant properties (Chen et al., 2022b; Sivieri et al., 2022). These beneficial dietary fibers are obtained from various sources, including cereals, mushrooms, seaweed, and fungi yeast. The physiological activity of beta-glucans is influenced by the type of linkage, whether branched or linear (beta-(1/3), (1/4), and (1/6)), connecting approximately 25,000 glucose monomers (Bobade et al., 2021; Hu et al., 2015; Tiwari & Cummins, 2009). The length and structure of BGs determine their functionality, which varies on considering the source, purification technique, and methods of extraction adopted. Usually, the complex structured BG exhibits enhanced biological activity than the simpler structure (Han et al., 2020; Lante et al., 2023; Mantovani et al., 2008). Insoluble fibre reduces transit time, adds on faecal mass, and improves the removal of bile salts. Soluble dietary fibres, on the other hand, are known for their prolonged satiety, delayed gastric emptying, and lower absorption rate (Bashir & Choi, 2017; Chanmuang et al., 2022; Daou & Zhang, 2012; Würsch & Pi-Sunyer, 1997).

Furthermore, the functionality of beta-glucans differs based on their source and structure. Bacterial BGs exhibit straight linear structure (β-1, 3 glucans) and improved gut health (Laroche & Michaud, 2008). Cereal-based BGs are linear as well, available as β-1, 3 and β-1, 4 links, protecting against cardiovascular disease (CVD) and other health-promoting properties, such as reducing cholesterol and improving overweight individual's insulin sensitivity (Chen & Raymond, 2008; Ho et al., 2016; Lin et al., 2013; Wolever et al., 2021). BG from mushrooms is linked to enhancing immune responses and possesses anticancer properties by stimulating immune cells and inhibiting tumor growth (Motta et al., 2021; Savelkoul et al., 2013; Steenwijk et al., 2021). Algae-derived BG has been shown to offer health advantages such as increasing immune cell activity, lowering inflammation, and improving the anti-bacterial effect against E.coli (Kadam et al., 2015; Kim et al., 2019). Fungal beta-glucans, specially yeast-derived are linked together by small β-1, 6 and branched β-1, 3 linkages and large β-1,6 and branched β-1,3 linkages, respectively, which accelerate and boost the host's immunity, protect upper respiratory tract against infections, improve gut health and reduce gut permeability, and inflammation (Steenwijk et al., 2021; Thomas et al., 2022).

## 1.7 Safety and Toxicity

It has been concluded from a large number of oral toxicity studies (Babícek et al., 2007; Jonker et al., 2010; Delaney et al., 2003a, b) that consumption of large concentration of beta-glucan did not cause any harmful effects. The results of genotoxicity studies (Chen et al., 2022a), as evaluated by gene mutations in Salmonella typhimurium, in vitro chromosome aberrations, and in vivo micronucleus test in mice, did not reveal any genotoxicity of H. erinaceus β-glucan extract preparation. Subchronic studies also revealed that the no-observed-adverse-effect level (NOAEL) for H. erinaceus β-glucan extract preparation was determined as 2000 mg/kg bw/day, the highest dose tested. The oral consumption of concentrated beta-glucan sourced from barley is not clastogenic (Delaney et al., 2004). The intake level of beta-glucan recommended by the FDA for lowering blood cholesterol is 50 mg/kg body weight/day (FDA, 2005). Beta-glucan administration at a concentration range of 5.8–5.9 g/kg body weight/day in Wistar rats showed no toxic symptoms (Jonker et al., 2010). It was found that administration of beta-glucan in concentrated form extracted from barley did not result in inflammatory or any other kind of consequences in CD-1 mice (Delaney et al., 2003b). It has also been reported that no adverse effects have been associated with salecan, a water-soluble β-1,3-glucan secreted by the Agrobacterium sp. ZX09, in subchronic and acute toxicity studies conducted on male and female mice. The salecan in mice has an LD50 value of above 3 g/kg for both male and female mice, as indicated by oral toxicity conducted with a single dose. The mice fed with salecan at a concentration of 5% or below level showed no adverse effects in the sub-chronic study. Salecan has not proven toxic at the tested level and displayed a no-observed-adverse-effect level (NOAEL) at 14.48 g/kg/day in mice (Bobade et al., 2022; Xiu et al., 2011).

## 1.8 Conclusion

β-Glucan is a soluble and non-starch homopolysaccharide. It is a dietary fiber primarily found in cereals, barley, seaweed, various bacteria, and fungi. β-glucan exhibits an excellent role in food processing due to its physicochemical and functional properties. It offers numerous health benefits, including reduction in blood glucose, cholesterol, obesity, anticancer, anti-inflammatory, and cardiovascular diseases. Oral toxicity studies in animals have confirmed it as safe at high doses of oral administration.

## References

Ahluwalia, B., & Ellis, E. E. (1984). A rapid and simple method for the determination of starch and β-glucan in barley and malt. *Journal of the Institute of Brewing, 90*(4), 254–259.

Ahmad, A., & Khalid, N. (2018). Dietary fibers in modern food production: A special perspective with β-glucans. *Biopolymers for Food Design*. https://doi.org/10.1016/B978-0-12-811449-0.00005-0

Ahmad, A., Anjum, F. M., Zahoor, T., Nawaz, H., & Din, A. (2009). Physicochemical and functional properties of barley β-glucan as affected by different extraction procedures. *International Journal of Food Science & Technology, 44*(1), 181–187.

Babícek, K., Cechová, I., Simon, R. R., Harwood, M., & Cox, D. J. (2007). Toxicological assessment of a particulate yeast (1,3/1,6)-b-d-glucan in rats. *Food and Chemical Toxicology*, 1719–1730. https://doi.org/10.1016/j.fct.2007.03.013

Babu, L. (2015). Green extraction techniques, structural analysis, and antioxidant activities of β-glucan present in oats. *International Journal of Latest Trends in Engineering and Technology, 5*, 125–135.

Bashir, K. M. I., & Choi, J. S. (2017). Clinical and physiological perspectives of β-glucans: The past, present, and future. *International Journal of Molecular Sciences, 18*(9). https://doi.org/10.3390/ijms18091906

Bhatty, R. S. (1995). Laboratory and pilot plant extraction and purification of β-glucans from hull-less barley and oat brans. *Journal of Cereal Science, 22*(2), 163–170. https://doi.org/10.1016/0733-5210(95)90046-2

Bobade, H., Gupta, A., & Sharma, S. (2021, January). Beta-glucan. In *Nutraceuticals and Health Care* (pp. 343–358). Elsevier. https://doi.org/10.1016/B978-0-323-89779-2.00013-2

Bobade, H., Gupta, A., & Sharma, S. (2022). Beta-glucan. In *Nutraceuticals and health care* (pp. 343–358). Academic.

Breedveld, M. W., & Miller, K. J. (1994). Cyclic beta-glucans of members of the family Rhizobiaceae. *Microbiological Reviews, 58*(2), 145–161.

Carr, J. M., Glatter, S., Jeraci, J. L., & Lewis, B. A. (1990). Enzymic determination of β-glucan in cereal-based food products. *Cereal Chemistry, 67*(3), 226–229.

Cerletti, C., Esposito, S., & Iacoviello, L. (2021). Edible mushrooms and beta-glucans: Impact on human health. *Nutrients, 13*(7). https://doi.org/10.3390/nu13072195

Chanmuang, S., Nguyen, Q. A., & Kim, H. J. (2022). Current research on the effects of non-digestible carbohydrates on metabolic disease. *Applied Sciences (Switzerland), 12*(8). https://doi.org/10.3390/app12083768

Chen, J., & Raymond, K. (2008). Beta-glucans in the treatment of diabetes and associated cardiovascular risks. *Vascular Health and Risk Management, 4*(6), 1265–1272. https://doi.org/10.2147/vhrm.s3803

Chen, S. N., Chang, C. S., Yang, M. F., Chen, S., Soni, M., & Mahadevan, B. (2022a). Subchronic toxicity and genotoxicity studies of Hericium erinaceus β-glucan extract preparation. *Current Research in Toxicology, 3*, 100068.

Chen, C., Huang, X., Wang, H., Geng, F., & Nie, S. (2022b). Effect of β-glucan on metabolic diseases: A review from the gut microbiota perspective. *Current Opinion in Food Science, 47*(Cvd), 100907. https://doi.org/10.1016/j.cofs.2022.100907

Choi, H., & Kim, S. W. (2023). Characterization of β-Glucans from cereal and microbial sources and their roles in feeds for intestinal health and growth of nursery pigs. *Animals, 13*(13). https://doi.org/10.3390/ani13132236

Daou, C., & Zhang, H. (2012). Oat Beta-Glucan: Its role in health promotion and prevention of diseases. *Comprehensive Reviews in Food Science and Food Safety, 11*(4), 355–365. https://doi.org/10.1111/j.1541-4337.2012.00189.x

Delaney, B., Carlson, T., Frazer, S., Zheng, T., Hess, R., Ostergren, K., Kierzek, K., Haworth, J., Knutson, N., Junker, K., & Jonker, D. (2003a). Evaluation of the toxicity of concentrated barley b-glucan in a 28-day feeding study in Wistar rats. *Food and Chemical Toxicology, 41*(4), 477–487. https://doi.org/10.1016/S0278-6915(02)00298-3

Delaney, B., Carlson, T., Zheng, G. H., Hess, R., Knutson, N., Frazer, S., Ostergren, K., Van Zijverden, M., Knippels, L., Jonker, D., & Penninks, A. (2003b). Repeated dose oral toxicological evaluation of concentrated barley b-glucan in CD-1 mice including a recovery phase. *Food and Chemical Toxicology, 41*(8), 1089–1102. https://doi.org/10.1016/S0278-6915(03)00046-2

Delaney, B., De Vogel, N., & Krul, C. A. M. (2004). Evaluation of the in vivo genetic toxicity of concentrated barley b-glucan. *Food and Chemical Toxicology, 42*(1), 155–156. https://doi.org/10.1016/S0278-6915(03)00219-9

FDA. (2005). Food labeling: Health claims; soluble dietaryfiber from certain foods and coronary heart disease.Interimfinal rule. *Federal Register, 70*, 76150–76162.

Han, B., Baruah, K., Cox, E., Vanrompay, D., & Bossier, P. (2020). Structure-functional activity relationship of β-glucans from the perspective of immunomodulation: A mini-review. *Frontiers in Immunology, 11*, 658.

Ho, H. V. T., Sievenpiper, J. L., Zurbau, A., Blanco Mejia, S., Jovanovski, E., Au-Yeung, F., Jenkins, A. L., & Vuksan, V. (2016). The effect of oat β-glucan on LDL-cholesterol, non-HDL-cholesterol and apoB for CVD risk reduction: A systematic review and meta-analysis of randomised-controlled trials. *British Journal of Nutrition, 116*(8), 1369–1382. https://doi.org/10.1017/S000711451600341X

Hu, X., Zhao, J., Zhao, Q., & Zheng, J. (2015). Structure and characteristic of β-Glucan in cereal: A review. *Journal of Food Processing and Preservation, 39*(6), 3145–3153. https://doi.org/10.1111/jfpp.12384

Irakli, M., Biliaderis, C. G., Izydorczyk, M. S., & Papadoyannis, I. N. (2004). Isolation, structural features and rheological properties of water-extractable β-glucans from different Greek barley cultivars. *Journal of the Science of Food and Agriculture, 84*(10), 1170–1178.

Islam, M. Z., An, H.-G., Kang, S.-J., & Lee, Y.-T. (2021). Physicochemical and bioactive properties of a high $β$-glucan barley variety 'Betaone' affected by germination processing. *International Journal of Biological Macromolecules, 177*, 129–134.

Izydorczyk, M. S., Storsley, J., Labossiere, D., MacGregor, A. W., & Rossnagel, B. G. (2000). Variation in total and soluble β-glucan content in hulless barley: Effects of thermal, physical, and enzymic treatments. *Journal of Agricultural and Food Chemistry, 48*(4), 982–989.

Jonker, D., Hasselwander, O., Tervilä-Wilo, A., & Tenning, P. P. (2010). 28-day oral toxicity study in rats with high purity barley beta-glucan (GlucagelTM). *Food and Chemical Toxicology*, 422–428. https://doi.org/10.1016/j.fct.2009.10.034

Kadam, S. U., Tiwari, B. K., & O'Donnell, C. P. (2015). Extraction, structure and biofunctional activities of laminarin from brown algae. *International Journal of Food Science and Technology, 50*(1), 24–31. https://doi.org/10.1111/ijfs.12692

Kaur, R., Sharma, M., Ji, D., Xu, M., & Agyei, D. (2019). Structural features, modification, and functionalities of beta-glucan. *Fibers, 8*(1), 1–29.

Kim, K., Ehrlich, A., Perng, V., Chase, J. A., Raybould, H., Li, X., Atwill, E. R., Whelan, R., Sokale, A., & Liu, Y. (2019). Algae-derived β-glucan enhanced gut health and immune responses of weaned pigs experimentally infected with a pathogenic E. coli. *Animal Feed Science and Technology, 248*(August 2018), 114–125. https://doi.org/10.1016/j.anifeedsci.2018.12.004

Lante, A., Canazza, E., & Tessari, P. (2023). Beta-glucans of cereals: Functional and technological properties. *Nutrients, 15*(9), 2124.

Laroche, C., & Michaud, P. (2008). New developments and prospective applications for β (1, 3) glucans. *Recent Patents on Biotechnology, 1*(1), 59–73. https://doi.org/10.2174/187220807779813938

Lazaridou, A., Biliaderis, C. G., & Izydorczyk, M. S. (2007). Cereal beta-glucans: structures, physical properties, and physiological functions. In *Functional food carbohydrates* (pp. 1–72). CRC Press.

Lin, N., Li, Y., Tang, L., Shi, J., & Chen, Y. (2013). In vivo effect of oat cereal β-glucan on metabolic indexes and satiety-related hormones in diet-induced obesity C57-Bl mice. *Molecular Nutrition and Food Research, 57*(7), 1291–1294. https://doi.org/10.1002/mnfr.201200695

Liu, H., Li, Y., You, M., & Liu, X. (2021). Comparison of physicochemical properties of β-glucans extracted from hull-less barley bran by different methods. *International Journal of Biological Macromolecules, 182*, 1192–1199.

Maheshwari, G., Sowrirajan, S., & Joseph, B. (2017). Extraction and isolation of β-glucan from grain sources—A review. *Journal of Food Science, 82*(7), 1535–1545.

Mantovani, M. S., Bellini, M. F., Angeli, J. P. F., Oliveira, R. J., Silva, A. F., & Ribeiro, L. R. (2008). β-Glucans in promoting health: Prevention against mutation and cancer. *Mutation Research – Reviews in Mutation Research, 658*(3), 154–161. https://doi.org/10.1016/j.mrrev.2007.07.002

Motta, F., Gershwin, M. E., & Selmi, C. (2021). Mushrooms and immunity. *Journal of Autoimmunity, 117*(November 2020), 102576. https://doi.org/10.1016/j.jaut.2020.102576

Muthuramalingam, K., Kim, Y., & Cho, M. (2021). β-glucan, "the knight of health sector": Critical insights on physiochemical heterogeneities, action mechanisms and health implications. *Critical Reviews in Food Science and Nutrition.*

Nakashima, A., Yamada, K., Iwata, O., Sugimoto, R., Atsuji, K., Ogawa, T., & Suzuki, K. (2018). β-Glucan in foods and its physiological functions. *Journal of Nutritional Science and Vitaminology, 64*(1), 8–17.

Nishantha, M. D. L. C., Zhao, X., Jeewani, D. C., Bian, J., Nie, X., & Weining, S. (2018). Direct comparison of $β$-glucan content in wild and cultivated barley. *International Journal of Food Properties, 21*(1), 2218–2228.

Papageorgiou, M., Lakhdara, N., Lazaridou, A., Biliaderis, C. G., & Izydorczyk, M. S. (2005). Water extractable (1→ 3, 1→ 4)-β-D-glucans from barley and oats: An intervarietal study on their structural features and rheological behaviour. *Journal of Cereal Science, 42*(2), 213–224.
Savelkoul, H. F. J., Chanput, W., & Wichers, H. J. (2013). Immunomodulatory effects of mushroom β-glucans. In *Diet, immunity and inflammation*. Woodhead Publishing Limited. https://doi.org/10.1533/9780857095749.3.416
Shehzad, A., Rabail, R., Munir, S., Jan, H., Fernández-Lázaro, D., & Aadil, R. M. (2023). Impact of oats on appetite hormones and body weight management: A review. *Current Nutrition Reports, 12*(1), 66–82. https://doi.org/10.1007/s13668-023-00454-3
Shoukat, M., & Sorrentino, A. (2021). Cereal β-glucan: A promising prebiotic polysaccharide and its impact on the gut health. *International Journal of Food Science and Technology, 56*(5), 2088–2097. https://doi.org/10.1111/ijfs.14971
Sivieri, K., de Oliveira, S. M., de Souza Marquez, A., Pérez-Jiménez, J., & Diniz, S. N. (2022). Insights on β-glucan as a prebiotic coadjuvant in the treatment of diabetes mellitus: A review. *Food Hydrocolloids for Health, 2*(January), 100056. https://doi.org/10.1016/j.fhfh.2022.100056
Skendi, A., Biliaderis, C. G., Lazaridou, A., & Izydorczyk, M. S. (2003). Structure and rheological properties of water soluble β-glucans from oat cultivars of Avena sativa and Avena bysantina. *Journal of Cereal Science, 38*(1), 15–31.
Suchecka, D., Gromadzka-Ostrowska, J., Zyła, E., Harasym, J. P., & Oczkowski, M. (2017). Selected physiological activities and health promoting properties of cereal beta-glucans. *A review. Journal of Animal and Feed Sciences, 26*(3).
Thomas, S., Rezoagli, E., Abidin, I. Z., Major, I., Murray, P., & Murphy, E. J. (2022). β-Glucans from Yeast—Immunomodulators from Novel Waste Resources. *Applied Sciences (Switzerland), 12*(10). https://doi.org/10.3390/app12105208
Tian, Y., Zeng, H., Xu, Z., Zheng, B., Lin, Y., Gan, C., & Lo, Y. M. (2012). Ultrasonic-assisted extraction and antioxidant activity of polysaccharides recovered from white button mushroom (Agaricus bisporus). *Carbohydrate Polymers, 88*(2), 522–529.
Tiwari, U., & Cummins, E. (2009). Factors influencing β-glucan levels and molecular weight in cereal-based products. *Cereal Chemistry, 86*(3), 290–301. https://doi.org/10.1094/CCHEM-86-3-0290
Van Steenwijk, H. P., Bast, A., & De Boer, A. (2021). Immunomodulating effects of fungal beta-glucans: From traditional use to medicine. *Nutrients, 13*(4), 1333.
Vannucci, L., Krizan, J., Sima, P., Stakheev, D., Caja, F., Rajsiglova, L., Horak, V., & Saieh, M. (2013). Immunostimulatory properties and antitumor activities of glucans. *International Journal of Oncology, 43*(2), 357–364.
Wolever, T., Rahn, M., Dioum, E., Spruill, S. E., Ezatagha, A., Campbell, J. E., Jenkins, A. L., & Chu, Y. (2021). An oat β-glucan beverage reduces LDL cholesterol and cardiovascular disease risk in men and women with borderline high cholesterol: A double-blind, randomized, controlled clinical trial. *Journal of Nutrition, 151*(9), 2655–2666. https://doi.org/10.1093/jn/nxab154
Würsch, P., & Pi-Sunyer, F. X. (1997). The role of viscous soluble fiber in the metabolic control of diabetes: A review with special emphasis on cereals rich in β-glucan. *Diabetes Care, 20*(11), 1774–1780. https://doi.org/10.2337/diacare.20.11.1774
Xiu, A., Zhan, Y., Zhou, M., Zhu, B., Wang, S., Jia, A., Dong, W., Cai, C., & Zhang, J. (2011). Results of a 90-day safety assessment study in mice fed a glucan produced by *agrobacterium* sp. ZX09. *Food and Chemical Toxicology, 49*(9), 2377–2384. https://doi.org/10.1016/j.fct.2011.06.050
Xu, C., Lv, J., Lo, Y. M., Cui, S. W., Hu, X., & Fan, M. (2013). Effects of oat β-glucan on endurance exercise and its anti-fatigue properties in trained rats. *Carbohydrate Polymers, 92*(2), 1159–1165.
Ye, S., Shah, B. R., Li, J., Liang, H., Zhan, F., Geng, F., & Li, B. (2022). A critical review on interplay between dietary fibers and gut microbiota. *Trends in Food Science & Technology, 124*, 237–249.
Zhu, F., Du, B., & Xu, B. (2016). A critical review on production and industrial applications of beta-glucans. *Food Hydrocolloids, 52*, 275–288. https://doi.org/10.1016/j.foodhyd.2015.07.003

# Chapter 2
# Sources of Beta-Glucan

**Shubham Rohilla, Akriti Jaiswal, Jinku Bora, and Maanas Sharma**

## 2.1 Introduction

β-Glucan (BG) is a polysaccharide which contains D-glucose monomer attached with other through glycosidic bonds. BG comes under the dietary fibre and can be consumed by the consumed from the different sources like cereal grains, edible mushrooms, microbial sources like bacteria, yeast, fungi etc. (Maheshwari et al., 2017). In cereals, the main type of fibre present in oats and barley is β-glucan and comes under the category of non-starch polysaccharides (Buckeridge et al., 2004). Glucan is the polysaccharides that contain glucose as main solo monomer unit in their structure. However, in their structure presence of glycogen, dextran, cellulose were present also and general molecular formula is $(C_6H_{12}O_5)_n$ (Rop et al., 2009). BG are composed of D-glucose monomers joined by glycosidic bonds and it is a long-chain. Cell walls from a wide variety of natural sources, including microorganism, yeasts like *Saccharomyces cerevisiae*, algae, edible mushrooms like *Ganoderma applanatum* and *reishi*, fungi like *Agaricus* and *Aspergillus* species and cereal grains including rye, oats,wheat, barley (Maheshwari et al., 2017). Though

S. Rohilla (✉)
Indian Institute of Packaging, New Delhi, India

A. Jaiswal
Indian Institute of Packaging, New Delhi, India

Department of Food Science and Technology, I.K Gujral Punjab Technical University, Punjab, India

J. Bora
Department of Food Technology, Jamia Hamdard, New Delhi, India

M. Sharma
Department of Food Technology and Home Science, Vivekananda College, University of Delhi, Vivek Vihar, Delhi, India

 
M. Bashir Mir, S. A. Mir (eds.), *Beta-Glucan: Sources, Properties and Applications*,
https://doi.org/10.1007/978-3-031-95788-8_2

microorganism are used industrially for the production of the BG and the most common microorganism include the category of yeast, mold etc. (Waszkiewicz-Robak, 2013). A small quantity of barley is utilised globally as a supply of β-glucan in a variety of meals for human consumption, although the majority of barley is utilised as animal feed (Ahmad et al., 2012). For normal metabolic activities, the organism body needs glucose with that these digestible polysaccharides provides glucose and calories as well. Apart from the energy, consuming these dietary fibre are known to promotes the body's wellness, but the extent of wellness depends upon the concentration and nature of source (Ferretti & Mariani, 2017). Li and Uppal (2010) reported that the recommended consumption of dietary fibre for men was in the range of 30–38 mg/day for men and for women it was 21–25 mg/day for better results. Consumption of dietary fibre improves the digestion process and helps in combat the treatment of various health diseases as these compounds are reported to maintain the homeostasis in the humans and act as pre-biotic compounds (Mejía et al., 2020). The Food Drug and Administration (FDA) claimed that β-glucan so toxic or side effects after consuming and considered to be in the category of GRAS (Generally Recognized as Safe) food category (Utama et al., 2020). The Codex Committee on Nutrition and Foods for Special Dietary Uses (CCNFSDU) reported that these dietary fibres are very helpful and include both the animal and edible plant sources, however the reported literature is available is from the plant sources (Ahmad et al., 2012). However, another agency named US Food and Drug Administration reported that regular consumption of 3 mg of the BG from the cereal sources along with the diet of low cholesterol helps the organism to reduce the risk of heart related diseases (Kaur et al., 2020). According to Triton Market Research, the β-glucan market, which includes the dairy industry as one of the branches that requires β-glucan as a technological ingredient, was estimated to be worth 340.63 million USD in 2018 and is expected to generate approximately 656.65 million USD in revenue by 2027 (Mykhalevych et al., 2022). Researchers have consistently demonstrated that polysaccharides, specifically β-glucan, are technologically useful components in food industry (Kamath et al., 2022).

Another group of researchers worked on the health benefits possess after consumption of BG includes activities like antioxidants, anti-tumour, anti-bacterial, wound healing, and lipid lowering (Du et al., 2014; Tiwari & Cummins, 2009). In 1980s, the first study was conducted to determine the antimicrobial potential of the β-glucan on pathogenic microbes *Staphylococcus aureus* and results shows positive results (Liang et al., 1998). Since then a number of study shows that β-glucan is safe to consume and can be used as an food ingredient for the devolvement of various food products (Mirończuk-Chodakowska et al., 2021). Similarly, as these BG are extensively used in the development and production of the food products as BG impart properties like thickening agent, fat substitute, emulsifying agents, film forming, water holding properties (Yuan et al., 2020). As these compounds have the ability to form the viscous compounds after coming in the contact in the large intestine with the water therefore, these compounds are associated with the health benefits. Therefore, this chapter aims to discuss the different sources directly or indirectly related to BG.

## 2.2 Different Sources of the β-Glucan

There are many sources for the production of β-Glucan (BG) such as cereals, yeast, bacteria, mold etc. The source and structure of the β-glucan has the main role in effecting their functional and nutritional quality. In literature, it was reported that total β-glucan content is higher in the cereal sources in comparison to the microbial sources (Choi & Kim, 2023). The digestibility of the cereal BG was reported to be in the range of 63–73%, however the total digestibility was found in the range of 89–93% (Choi & Kim, 2023). Similarly, the β-glucan obtained from the microbial sources (*Saccharomyces cerevisiae*) shows 53–83% of insoluble fraction and main reason behind the difference in the range is the structural difference of bonds in the β-glucan (Suchecka et al., 2017). However, in the other non-cereal samples like yeast, bacteria, algae and mushrooms, β-Glucan are present in the inside the sample matric in the form of fibrous structure (Ahmad & Khalid, 2018). Apart from the fibrous structures, these compounds are linked themselves in the linkage of 1–3 β-glycosidic and 1–6 β-glycosidic linkages which is different from the linkage present in the cereal grains. Glucans are the main components that hold the major share in constituting the cell wall of the fungi and yeast, however in the yeast it is mushrooms it is one of the main polysaccharide (Cerletti et al., 2021). The difference among the linkage of the cereal and non-cereal samples is main factor that affects the quality of the BG, similarly the different properties were observed by the BG in terms of chemical, molecular and functional characteristics like solubility, digestibility were influenced greatly (Zhu et al., 2016). On the other hand factors like primary structure, polymer charge, solubility, degree of branching, molecular weight plays important role in providing the biological functional values of the β-glucan (Muthuramalingam et al., 2021; Shoukat & Sorrentino, 2021). The sources and the functional properties of β-glucan with the types of bonds is depicted in Table 2.1.

## 2.3 β-Glucan from Cereals

Humans have discovered different sources for the extraction as well as production of β-Glucan (BG), however the cereal grains are known to be the most vital sources for the obtainment of the BG. β-glucan is a heterogeneous group comprised of D-glucose polymers, known to exhibit diverse range of properties because of their diverse morphology. Cereals are composed of the fibrous structures that are attached with the help of the mixed linkages. In scientific literature, it was reported that cereals contains many different types of the dietary fibres, but β-glucans (BG) have received the greatest research attention because of their exceptional functional, physiological and technological properties (Schmidt, 2022). The linkages present in the glucan are formed with the combination of 1–3 β-glcosisidic and 1–4 β-glcosisidic bonds (Mejía et al., 2020). In the endosperm of some cereals, including oats, barley,

**Table 2.1** Different sources of the β-glucan with their functional properties (Lante et al., 2023)

| Sources | Species | Name of the β-glucan | Solubility in water | Glycosidic bonds |
|---|---|---|---|---|
| **Cereals** | | | | |
| Oats | *Avena sativa* | – | Soluble | β-1.3 and β-1.4 w/o branches |
| Barley | *Hordeum Vulgar* | – | Soluble | |
| **Fungi** | | | | |
| Schizophyllan | *Schizophyllum commune* | GSP | Soluble | Main chain consists of β-1,3 glycosidic bonds with branches with single glucose units β-1,6 |
| Maitake mushroom | *Leafy griffon* | Grifolan | Soluble | |
| Shiitake mushroom | *Lentinula edodes* | Lentinan | Soluble | |
| **Seaweeds** | | | | |
| Algae | *Laminaria typed* | Laminarin | Soluble | Main chain consisting of β-1,3 glycosidic bonds with β-1,6 side chains |
| Cochayuyo | *Durvillaea antarctica* | Laminarin | Soluble | |
| Algae | *Saccharina longicruris* | Laminarin | Soluble | |
| Microalgae | *Euglena gracilis* | Paramylon | Insoluble | Branchless chains β-1.3 |
| **Yeast** | | | | |
| Brewer's yeast | *Saccharomyces cerevisiae* | Zymosan | Insoluble | Main chain consisting of β-1,3 glycosidic bonds with β-1,6 side chains |

wheat, and rice, β-glucan with these structures are reported to found in the aleurone, the subaleurone layer, and and the cell walls (Mejía et al., 2020). In cereal grains such as oats, barley, rice and wheat, β-glucan are present in mostly located in the aleurone layer of the sample matrix. The pattern present in the cereal grain consist of the β-D-Glucose linked with the consecutive linkages of β-(1 → 4) linked in the blocks and separated by a β-(1 → 3) linkage, build a structure which look like the staircase. The polysaccharide is built mainly of β-(1 → 3) linked cellotriosyl (58–72%) with the addition of cellotetraosyl (20–34%) units (Benito-Román et al., 2013; Tosh et al., 2004). The difference in the structures in the BG obtained from the different cereals are identified by the ratio of trisaccharide to tetrasaccharide, which varies from cereals to cereals (Henrion et al., 2019). The concentration of BG is varied depending upon the species of cereals such as greater amount of BG were present in barley sources followed by oat and least amount were present in rice and wheat (Mejía et al., 2020).

β-glucan is a linear homopolymer of D-glucopyranose units, which are composed of two or three consecutive β-(1–4) linkages separated by β-(1–3) linkages with the degree of polymerization of 5 to 28. The 'β' configuration of the β-glucan is not enzyme friendly means it is not digestible by the enzymes in human gut (Mejía et al., 2020), and the linkages of β-(1–3), make the BG molecules mores soluble, therefore because of these molecule falls under the category of soluble dietary fibre (Lazaridou et al., 2004). The content of BG is low in cereals ranging

**Table 2.2** Characteristics of β-glucan isolated from different cereals (Schmidt, 2022)

| Cereals | Molar ratio (DP3/DP4) | β-glucan content (%) | Molar mass (kg/mol) | Protein (%) | Starch (%) | Pentosans (%) |
|---|---|---|---|---|---|---|
| **Wheat** | 3.0–4.5 | ≈0.5 | 209–416 | 14 | 70 | 2–6 |
| **Rye** | 2.4–2.7 | ≈0.2 | Up to 1130 | 12 | 65 | 10 |
| **Barley** | 2.3–3.4 | 5.0–11.0 | 400–2500 | 11 | 64 | Less amount |
| **Oats** | 3.0–7.0 | 3.0–7.0 | 400–2500 | 12 | 59 | Less amount |

from 4% to 7% and highest molecular weight of BG is observed in oats (3,000,000 Da) and in barley (2,100,000 Da) (Henrion et al., 2019). The high molecular weight of the BG, which is present in the cereals and cereals based products, possess to have more impact on reducing the glycaemic index, postprandial blood glucose response (PBGR) (Henrion et al., 2019). Another group of researchers reported that molar mass of the BG obtained from rye, wheat, oats and barley was found in the range of 21–1.100, 209–416, 35–3.100 and 31–2.700 × $10^3$ g/mol, respectively (Lazaridou & Biliaderis, 2007). The BG obtained from different cereals shows the similar structure however the difference is on the basis of DP3 (cellotriose) /DP4 (cellotetraose) ratio. Apart from the ratio, the depolymerisation of the β-glucan polymer chain also play the major role in difference in the structure of the BG (Lante et al., 2023). A group of researcher reported that BG present in pearling by-products possess more (DP3/DP4) molar ratio (~3.3–3.6) and in fibre (Izydorczyk & Dexter, 2008; Schulze et al., 2007). The molar ratio (DP3/DP4) and percentage of β-glucan content present in the different cereals are depicted in Table 2.2. The functional properties exhibited by the β-glucan like gelation solubility, dispersibilty, viscosity etc. are mainly greatly affected by the molecular weight of the β-glucan (Mathews et al., 2021; Zhu et al., 2016). β-glucans physicochemical properties are depending upon the attributes of their main structure, such as the type of linkage, degree of branching, molecular weight, and conformation (such as single, triple, and random coil configurations) (Kofuji et al., 2012; Kwong et al., 2013).

## 2.4 β-Glucan from the Mycelium and Fruiting Body of Fungi

The production of β-glucan from fungal source gained the interest of researchers because of their potential health benefits and their exceptional functional properties (Zhu et al., 2016). The structure of BG has the common backbone β(1,3)-linked β-D-glucopyranosyl units, though the length and branching structure was found to be different in the fungal and cereals sources (Camilli et al., 2018). The most prevalent polysaccharide in fungal cell walls, β-glucan are primarily identified by the existence of β-(1,6)-linked branches originating from the β(1,3) backbone. On the other hand, fungal source also affects the structural diversity of the BG (Camilli et al., 2018; Synytsya & Novák, 2013). In yeast, the structure has β(1,6)-side branches with additional β(1,3) areas, however, the β-glucan from mushrooms contain short β(1,6)-linked branches (Borchers et al., 2004). Notably, various researches reported that a higher level of structural complexity is linked to improved β-glucan-induced properties like antibacterial and anticancer activities. Though, these structural variations may also affect the immunogenic qualities of β-glucan (Camilli et al., 2018; Chen & Seviour, 2007). The fungal glucan are divided into two major classes i.e. alpha bound glucan and beta bound glucan. These glucan shows different properties like alpha bound glucan are amorphous and soluble in hot water and act as energy reserve materials like starch in algae and glycogen in fungi. On the other hand, the beta bound glucan are crystalline and not soluble in water (Du et al., 2014). The further classification of the alpha glucan was divided into following two classes: (i) linear α-D-glucan and (ii) Branched α-D-glucan (Ruiz-Herrera & Ortiz-Castellanos, 2019). Similarly to the alpha glucan, the beta glucan was also reported into two categories: (i) linear β-D-glucan, (ii) branched β-D-glucan and (iii) contains both α and β bonds (Ruiz-Herrera & Ortiz-Castellanos, 2019). Researcher reported that fungi has more potential of production of β-glucan in comparison to the yeast (Bangar & Siroha, 2022; Mirończuk-Chodakowska et al., 2021). Apart from the BG present in the mushroom, as mentioned in Fig. 2.1, BG from these category are known to possess various health beneficial properties such as hypotensive, antioxidant, hypocholesterolemic, anticancer, antiviral, bacteriostatic properties and recommended as food and food products (Zhang et al., 2017). The most common method of extraction of BG from the fungi was done using hot boiling water. Researchers found that two water insoluble glucans were isolated from the fruiting bodies of Mushroom (*Ramaria botrytis*) (Bhanja et al., 2014). Kim et al. (2005) extracted the β-glucan from the fruting bodies from *Agaricus blazei* using hot water extraction and purified using the column named Sephadex G-50 and Sephadex A-25.

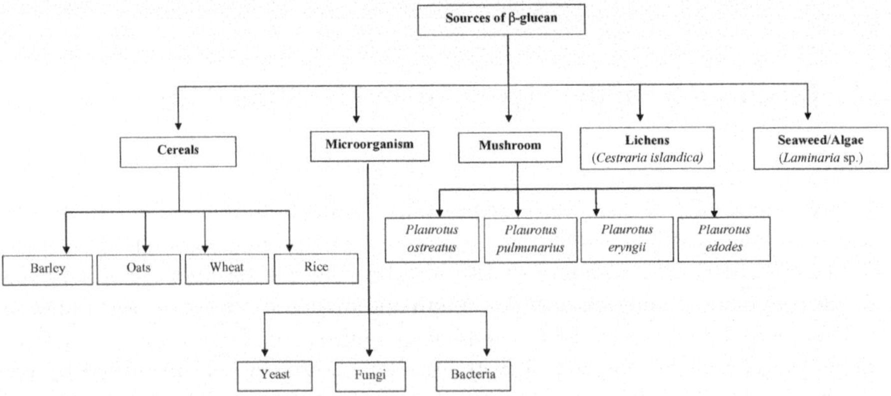


**Fig. 2.1** Different sources of *β-glucan* (Kaur et al., 2020)

## 2.5 β-Glucan from Yeast

In yeast, the β-Glucan (BG) is exists inside the inner layer of the cell of the yeast and the structure of is composed of monosaccharide units of glucose with the presence of β-(1 → 3)-glucan with the attaches side-chain of β- (1 → 6)-glucan, in which β-(1 → 3)-glucan is connected by β-(1 → 3) glycosidic bonds and β-(1 → 6)-glucan is connected by β-(β-Glucan (BG)1 → 6) glycosidic bonds (Liu et al., 2021; Yiannikouris et al., 2004). β-glucan is the principal component of the yeast cell wall with the range of 29–64%, however apart from that the concentration of other components accounts is protein 13%, mannan 31%, lipids 9% and chitin 1–2%, respectively (Bacon et al., 1969). Thus, mannan and BG are considered to be main active components a.k.a. manno-proteins of the yeast cell and these manno-proteins are known to possess various health benefits properties (Liu et al., 2021).One of the researcher reported similar work that *S.cerevaise* cell wall structure contains protein bound to sugar known as mannoprotein, which helps to give organism structure and act as food reserve (Ruiz-Herrera & Ortiz-Castellanos, 2019). In western countries, a food that contains BG are consumed for thousands of years. These foods are more referred as dietary supplements are mainly derived from the baker's yeast (*Saccharomyces cerevisiae*) (Legras et al., 2007). *Saccharomyces cerevisiae* is unicellular yeast was found to be the most effective and widely used yeast, utilize for synthesizing the β-glucans on large scale. It is a non-toxic and non-pathogenic microorganism (Du et al., 2014). The microorganisms used for the production of the β-glucan are presented in Fig. 2.1. The extracted β-glucan from the baker's yeast is considered to be possess various health benefits and reported to lowering the cholesterol and helps to enhance to improve the immunity of the consumers (Ahmad & Kaleem, 2018). Research also shows that consumption of yeast β-glucan is the potential candidate that has the tendency to enhance the intestinal health and shows reduction of the pathogenic microorganism in the pig gut (Choi & Kim, 2023). Author studied the antioxidant and anti-inflammatory activities of β-glucan extracted from Mushroom (*Geastrum saccatum*) and found that positive results on superoxide radical, lipid peroxidation, hydroxyl radical assays (Guerra Dore et al., 2007). Method of cultivation of yeast has a prominent effect on the glucan content, however, the type of branching and bonds decide which type of glucan is present inside the cell (Boutros et al., 2022).

## 2.6 β-Glucan from Bacteria

Bacterial β-glucan (BG) are reported to be produced using multi-stage process that required the growth of mycelium, substrate, and catalyzation of glucan-hydrolytic enzymes in the bioreactor (Zhu et al., 2016). Crognale et al. (2007) studied the production of β-glucan from *Botryosphaeria rhodina* DABAC-P82 using the required conditions such as oxygen transfer, culture medium rheology etc. They found and reported that maximum amount of β-glucan was produced in the bio-reactor and the factors that are associated with them are nitrogen and oxygen. Another group of researcher studied the production of β-glucan from the *Lactobacillus paracasei* using the membrane associated enzymes and reported their results produced β-glucan possess very a good potential in food as well as pharmaceutical industry (Zhu et al., 2016). The structure of bacterial β-glucan produced by *Elsinoe leucospila* consist of β-(1 → 3) glycosidic linkages, some other β-glucan consist of β-(1 → 4) glycosidic linkages (Ahmad et al., 2012; Chen et al., 2022). On the other hand, bacterial BG such as *Agribacterium biobaris* have a straight and unbranched β (1 → 3)(1 → 3)-D-glucan backbones and the structure possessed by seaweeds such as *luminaria* is composed of β (1 → 3)(1 → 3) (Bonfim-Mendonça et al., 2017). *Xanthomonas campestris* is a gram negative bacteria which are known to produce the β-(1,2) glucan and *Bacillus natto* produce the β-glucanase enzymes (Moriartey et al., 2010). The side branch chain β-(1,3;1,2)-D-glucan is reported to be present in the BG extracted from bacterium *Streptococcus pneumonia* (McIntosh et al., 2005).

## 2.7 β-Glucan from Other Sources

There are different sources of the β-glucan (BG) which include plants, microbes etc. Nowadays, researchers also more focused on the non-conventional source of the β-glucan. Yoshida et al. (2014) produced the BG from the corn pericarp and extracted using the alkali solutions along with the purification was done using the affinity and anion-exchange chromatography. Similarly, a another group of researcher from Japan reported that Japanese cedar bark was used for the production of BG A group of researchers extracted the BG from the six different Korean alcoholic beverage (*Makgeolli*) and reported that R-12 commercial raw *Makgeolli* possess high content β-glucan with additional immune-modulating property (Min et al., 2012). Marine brown algae *Laminaria digitata* was used for the production of BG with the combination of two step extraction namely hot water extraction and ultrafiltration. The purification of the BG was done using the phycarine (Vetvicka et al., 2007). Watanabe et al. (2013) studied the BG extracted from the Euglena and found that it shows antitumor activity against preneoplastic colonic aberrant crypt foci in mice. The other species that are known for producing the BG includes *Rhizobium sp., R. leguminosarum, R. meliloti, A. tumefaciens* (Breedveld & Miller, 1994). According to the Ahmad and Kaleem (2018) other non-conventional sources that are reported to produce the BG are present in the beans, millet (*Panicum miliaceum*), lentils (*Lens culinaris*) and canary seeds (*Tropaeolum peregrinum*).

## 2.8 Solubility of β-Glucan

The solubility of the β-glucan (BG) is an important parameter that affects physicochemical and physiological functional properties. The molecular irregularity is present in the BG, which is mainly responsible for affecting the solubility, though, most of the BG sources is water soluble as mentioned in Table 2.1 (El Khoury et al., 2012). However, the main reason behind the higher solubility of β-glucan than cellulose is due to the presence of irregular β-(1–3) bonds in the structure of BG (Schmidt, 2022). The extraction method also play important role in the solubilisation power of the BG, mild extraction possess lower solubility and vigorous conditions possess higher solubility (Mathews et al., 2021). The parameters like subsequent extraction, pH and other physical extraction conditions play important role in affecting the solubility of β-glucan. Though, BG extracted from the oats shows higher solubility in comparison to BG from the barley, and the reason behind stated is the presence of more molar ratio which lowers the solubility (Lazaridou & Biliaderis, 2007). Nonetheless, there is no clear separation or distinction has been reported so far on the soluble and insoluble fractions, but the ratio is mainly responsible on the solubility (Virkki et al., 2005). A another group of researchers reported that interaction between the components of the cell and spatial organization also responsible for the mechanical strength and permeability which affecting the solubility of the compounds (Izydorczyk & Dexter, 2008). It was also reported that this limited increase in the depolymerisation of the polysaccharide enhances the solubility and stability of the solutions (Izydorczyk & Dexter, 2008). Beta glucan from the oats shows higher solubility & stability followed by barley and least was shown by wheat (Lante et al., 2023). The presence of the $SO_2$ and COOH group helps in enhancing the solubility of the polysaccharides (Elleuch et al., 2011). The branching in the β-glucan contains (1 → 6) linkage in the bacterial and fungal sources, thus this branching is related to the produce mostly insoluble BG which are known are related to the activation/inactivation of specific receptors (Nakashima et al., 2018).

## 2.9 Extraction of β-Glucan

The extraction of the β-glucan (BG) is the most important and crucial step to extract the produced BG from the different sources, and for this different extraction methods and purification steps has been employed for the production of pure BGs (Zhu et al., 2016). However, the source of the extraction matters a lot for extracting the BGs such as in BG extracted from fungal source required a number of technical steps with the sound knowledge of fermentation (Crognale et al., 2007). On the other hand, the molecular weight and structure affect nature of the extraction procedure and yield of the BG. But in booth the cases, a range of extraction as well as purification methods are available for BG extraction (Ahmad et al., 2012; Zhu et al., 2016).The downstream process of β-glucan was reported to divided into main categories following: wet and dry extraction conditions whereas, careful monitoring and adjustment procedure was taken care of while performing the extraction procedure for the obtainment of better results (Benito-Román et al., 2011). In scientific literature, it was mentioned that extraction of β-glucan from cereal sources are done mainly divided into four categories following: (i) acidic extraction, (ii) enzymatic extraction (iii) water extraction and (iv) alkaline extraction (Maheshwari et al., 2017). A another group of researcher studied the extraction of BG from the cereals sources becomes more difficult as well as expensive in comparison to other sources of BG (Daou & Zhang, 2012). In literature, it was reported that beta glucan extracted from cereal using mild reagents and conditions yield incomplete extraction. To overcome this problem, researchers used vigorous conditions during extraction such as alkaline conditions where yield is reported to be improved because this extraction helps in depolymerize the polysaccharide chain to give smaller molecular weight beta glucan (Beer et al., 1997). Many different types of extraction were employed in the extraction of BG, however the most common, easy, economic and widely employed extraction method reported in the scientific literature is hot water extraction (Zhu et al., 2016). The other methods include the alkali extraction methods, accelerated water extraction, ultrasound assisted extraction, acidic extraction, microwave-assisted extraction methods reported so far used to extract the BG, however, these methods are known for better yields but the drawback they are carrying is the running cost and reduced environment sustainability and long running time (Routray & Orsat, 2012; Zhu et al., 2016). After extraction of the BG, the main step is to purify the BG and separate it from the starch and protein, however this step is optional in nature and not followed. Researchers recommended that using enzymatic treatment with α-amylase and α-protease is known for giving best purified BG (Limberger-Bayer et al., 2014; Mejía et al., 2020).

## 2.10 Conclusion

β-Glucan (BG) has a wide range of applications and health advantages because it comes from different sources such bacteria, fungi, yeast, and algae. Being mostly composed of D-glucose polymers attached by glycosidic linkages, BG is an essential component of dietary fibre and provides a number of health benefits. Its occurrence in microbial sources and fungi, as well as in cereals like oats and barley, indicates that BG has a variety of structural complexity and qualities that affect its solubility, digestibility, and physiological effects. Given its functional and therapeutic qualities, beta-glucan is one of the polysaccharides that is most well-known. Naturally, β-glucan is present in a wide variety of foods such as plants especially oat and barley (cereals). Moreover, some microorganisms, such as yeasts, bacteria are also known to be used as the sources for the production of β-glucan. Different sources of β-glucan exhibit different structural characteristics and each have different functional characteristics.

**Acknowledgement** Shubham Rohilla is thankful to IKGPTU for providing necessary facilities.

## References

Ahmad, A., & Kaleem, M. (2018). β-glucan as a food ingredient. In *Biopolymers for food design*. Elsevier. https://doi.org/10.1016/B978-0-12-811449-0.00011-6

Ahmad, A., & Khalid, N. (2018). Dietary fibers in modern food production: A special perspective with β-glucans. In *Biopolymers for food design*. Elsevier. https://doi.org/10.1016/B978-0-12-811449-0.00005-0

Ahmad, A., Anjum, F. M., Zahoor, T., Nawaz, H., & Dilshad, S. M. R. (2012). Beta glucan: A valuable functional ingredient in foods. *Critical Reviews in Food Science and Nutrition, 52*(3). https://doi.org/10.1080/10408398.2010.499806

Bacon, J. S., Farmer, V. C., Jones, D., & Taylor, I. F. (1969). The glucan components of the cell wall of baker's yeast (Saccharomyces cerevisiae) considered in relation to its ultrastructure. *The Biochemical Journal, 114*(3), 557–567. https://doi.org/10.1042/bj1140557

Bangar, S. P., & Siroha, A. K. (2022). Functional cereals and cereal foods: Properties, functionality and applications. In *Functional cereals and cereal foods: Properties, functionality and applications*. https://doi.org/10.1007/978-3-031-05611-6

Beer, M. U., Wood, P. J., & Weisz, J. (1997). Molecular weight distribution and (1→3)(1→4)-β-D-glucan content of consecutive extracts of various oat and barley cultivars. *Cereal Chemistry, 74*(4). https://doi.org/10.1094/CCHEM.1997.74.4.476

Benito-Román, O., Alonso, E., & Lucas, S. (2011). Optimization of the β-glucan extraction conditions from different waxy barley cultivars. *Journal of Cereal Science, 53*(3), 271–276. https://doi.org/10.1016/j.jcs.2011.01.003

Benito-Román, O., Alonso, E., & Cocero, M. J. (2013). Pressurized hot water extraction of B-glucans from waxy barley. *Journal of Supercritical Fluids, 73*, 120–125. https://doi.org/10.1016/j.supflu.2012.09.014

Bhanja, S. K., Rout, D., Patra, P., Sen, I. K., Nandan, C. K., & Islam, S. S. (2014). Water-insoluble glucans from the edible fungus Ramaria botrytis. *Bioactive Carbohydrates and Dietary Fibre, 3*(2), 52–58. https://doi.org/10.1016/j.bcdf.2014.01.004

Bonfim-Mendonça, P. S., Capoci, I. R. G., Tobaldini-Valerio, F. K., Negri, M., & Svidzinski, T. I. E. (2017). Overview of β-glucans from laminaria spp.: Immunomodulation properties and applications on biologic models. *International Journal of Molecular Sciences, 18*(9). https://doi.org/10.3390/ijms18091629

Borchers, A. T., Keen, C. L., & Gershwini, M. E. (2004). Mushrooms, tumors, and immunity: An update. *Experimental Biology and Medicine, 229*(5). https://doi.org/10.1177/153537020422900507

Boutros, J. A., Magee, A. S., & Cox, D. (2022). Comparison of structural differences between yeast β-glucan sourced from different strains of saccharomyces cerevisiae and processed using proprietary manufacturing processes. *Food Chemistry, 367*, 130708. https://doi.org/10.1016/j.foodchem.2021.130708

Breedveld, M. W., & Miller, K. J. (1994). Cyclic β-glucans of members of the family Rhizobiaceae. *Microbiological Reviews, 58*(2), 145–161. https://doi.org/10.1128/mmbr.58.2.145-161.1994

Buckeridge, M. S., Rayon, C., Urbanowicz, B., Tiné, M. A. S., & Carpita, N. C. (2004). Mixed Linkage (1→3),(1→4)-β-D-Glucans of Grasses. *Cereal Chemistry, 81*(1). https://doi.org/10.1094/CCHEM.2004.81.1.115

Camilli, G., Tabouret, G., & Quintin, J. (2018). The complexity of fungal β-Glucan in health and disease: Effects on the mononuclear phagocyte system. *Frontiers in Immunology, 9*(APR). https://doi.org/10.3389/fimmu.2018.00673

Cerletti, C., Esposito, S., & Iacoviello, L. (2021). Edible mushrooms and beta-glucans: Impact on human health. *Nutrients, 13*(7). https://doi.org/10.3390/nu13072195

Chen, J., & Seviour, R. (2007). Medicinal importance of fungal β-(1→3), (1→6)-glucans. *Mycological Research, 111*(6). https://doi.org/10.1016/j.mycres.2007.02.011

Chen, C., Huang, X., Wang, H., Geng, F., & Nie, S. (2022). Effect of β-glucan on metabolic diseases: A review from the gut microbiota perspective. *Current Opinion in Food Science, 47*, 100907. https://doi.org/10.1016/j.cofs.2022.100907

Choi, H., & Kim, S. W. (2023). Characterization of β-glucans from cereal and microbial sources and their roles in feeds for intestinal health and growth of nursery pigs. *Animals, 13*(13). https://doi.org/10.3390/ani13132236

Crognale, S., Bruno, M., Fidaleo, M., Moresi, M., & Petruccioli, M. (2007). Production of β-glucan and related glucan-hydrolases by Botryosphaeria rhodina. *Journal of Applied Microbiology, 102*(3), 860–871. https://doi.org/10.1111/j.1365-2672.2006.03116.x

Daou, C., & Zhang, H. (2012). Oat Beta-Glucan: Its role in health promotion and prevention of diseases. *Comprehensive Reviews in Food Science and Food Safety, 11*(4), 355–365. https://doi.org/10.1111/j.1541-4337.2012.00189.x

Du, B., Bian, Z., & Xu, B. (2014). Skin health promotion effects of natural Beta-Glucan derived from cereals and microorganisms: A review. *Phytotherapy Research, 28*(2). https://doi.org/10.1002/ptr.4963

El Khoury, D., Cuda, C., Luhovyy, B. L., & Anderson, G. H. (2012). Beta glucan: Health benefits in obesity and metabolic syndrome. In. *Journal of Nutrition and Metabolism, 2012*, 1–28. https://doi.org/10.1155/2012/851362

Elleuch, M., Bedigian, D., Roiseux, O., Besbes, S., Blecker, C., & Attia, H. (2011). Dietary fibre and fibre-rich by-products of food processing: Characterisation, technological functionality and commercial applications: A review. *Food Chemistry, 124*(2). https://doi.org/10.1016/j.foodchem.2010.06.077

Ferretti, F., & Mariani, M. (2017). Simple vs. complex carbohydrate dietary patterns and the global overweight and obesity pandemic. *International Journal of Environmental Research and Public Health, 14*(10). https://doi.org/10.3390/ijerph14101174

Guerra Dore, C. M. P., Azevedo, T. C. G., de Souza, M. C. R., Rego, L. A., de Dantas, J. C. M., Silva, F. R. F., Rocha, H. A. O., Baseia, I. G., & Leite, E. L. (2007). Antiinflammatory, antioxidant and cytotoxic actions of β-glucan-rich extract from Geastrum saccatum mushroom. *International Immunopharmacology, 7*(9), 1160–1169. https://doi.org/10.1016/j.intimp.2007.04.010

Henrion, M., Francey, C., Lê, K. A., & Lamothe, L. (2019). Cereal B-glucans: The impact of processing and how it affects physiological responses. *Nutrients, 11*(8). https://doi.org/10.3390/nu11081729

Izydorczyk, M. S., & Dexter, J. E. (2008). Barley β-glucans and arabinoxylans: Molecular structure, physicochemical properties, and uses in food products-a review. *Food Research International, 41*(9), 850–868. https://doi.org/10.1016/j.foodres.2008.04.001

Kamath, R., Basak, S., & Gokhale, J. (2022). Recent trends in the development of healthy and functional cheese analogues-a review. *LWT, 155*, 112991. https://doi.org/10.1016/j.lwt.2021.112991

Kaur, R., Sharma, M., Ji, D., Xu, M., & Agyei, D. (2020). Structural features, modification, and functionalities of beta-glucan. *Fibers, 8*(1). https://doi.org/10.3390/fib8010001

Kim, Y. W., Kim, K. H., Choi, H. J., & Lee, D. S. (2005). Anti-diabetic activity of β-glucans and their enzymatically hydrolyzed oligosaccharides from Agaricus blazei. *Biotechnology Letters, 27*(7), 483–487. https://doi.org/10.1007/s10529-005-2225-8

Kofuji, K., Aoki, A., Tsubaki, K., Konishi, M., Isobe, T., & Murata, Y. (2012). Antioxidant activity of β-glucan. *ISRN Pharmaceutics, 2012*, 1–5. https://doi.org/10.5402/2012/125864

Kwong, M. G. Y., Wolever, T. M. S., Brummer, Y., & Tosh, S. M. (2013). Increasing the viscosity of oat β-glucan beverages by reducing solution volume does not reduce glycaemic responses. *British Journal of Nutrition, 110*(8), 1465–1471. https://doi.org/10.1017/S000711451300069X

Lante, A., Canazza, E., & Tessari, P. (2023). Beta-glucans of cereals: Functional and technological properties. *Nutrients, 15*(9). https://doi.org/10.3390/nu15092124

Lazaridou, A., & Biliaderis, C. G. (2007). Molecular aspects of cereal β-glucan functionality: Physical properties, technological applications and physiological effects. *Journal of Cereal Science, 46*(2), 101–118. https://doi.org/10.1016/j.jcs.2007.05.003

Lazaridou, A., Biliaderis, C. G., Micha-Screttas, M., & Steele, B. R. (2004). A comparative study on structure-function relations of mixed-linkage (1→3), (1→4) linear β-D-glucans. *Food Hydrocolloids, 18*(5), 837–855. https://doi.org/10.1016/j.foodhyd.2004.01.002

Legras, J. L., Merdinoglu, D., Cornuet, J. M., & Karst, F. (2007). Bread, beer and wine: Saccharomyces cerevisiae diversity reflects human history. *Molecular Ecology, 16*(10), 2091–2102. https://doi.org/10.1111/j.1365-294X.2007.03266.x

Li, C., & Uppal, M. (2010). Canadian diabetes association national nutrition committee clinical update on dietary fibre in diabetes: Food sources to physiological effects. *Canadian Journal of Diabetes, 34*(4), 355–361. https://doi.org/10.1016/S1499-2671(10)44010-1

Liang, J., Melican, D., Cafro, L., Palace, G., Fisette, L., Armstrong, R., & Patchen, M. L. (1998). Enhanced clearance of a multiple antibiotic resistant Staphylococcus aureus in rats treated with PGG-glucan is associated with increased leukocyte counts and increased neutrophil oxidative burst activity. *International Journal of Immunopharmacology, 20*(11), 595–614. https://doi.org/10.1016/S0192-0561(98)00007-1

Limberger-Bayer, V. M., De Francisco, A., Chan, A., Oro, T., Ogliari, P. J., & Barreto, P. L. M. (2014). Barley β-glucans extraction and partial characterization. *Food Chemistry, 154*, 84–89. https://doi.org/10.1016/j.foodchem.2013.12.104

Liu, Y., Wu, Q., Wu, X., Algharib, S. A., Gong, F., Hu, J., Luo, W., Zhou, M., Pan, Y., Yan, Y. Y., & Wang, Y. (2021). Structure, preparation, modification, and bioactivities of β-glucan and mannan from yeast cell wall: A review. *International Journal of Biological Macromolecules, 173*, 445–456. https://doi.org/10.1016/j.ijbiomac.2021.01.125

Maheshwari, G., Sowrirajan, S., & Joseph, B. (2017). Extraction and isolation of β-glucan from grain sources—A review. *Journal of Food Science, 82*(7), 1535–1545. https://doi.org/10.1111/1750-3841.13765

Mathews, R., Kamil, A., & Chu, Y. F. (2021). Global review of heart health claims for oat beta-glucan products. *Nutrition Reviews, 78*, 78–97. https://doi.org/10.1093/NUTRIT/NUZ069

McIntosh, M., Stone, B. A., & Stanisich, V. A. (2005). Curdlan and other bacterial (1→3)-β-D-glucans. *Applied Microbiology and Biotechnology, 68*(2). https://doi.org/10.1007/s00253-005-1959-5

Mejía, S. M. V., de Francisco, A., & Bohrer, B. M. (2020). A comprehensive review on cereal β-glucan: Extraction, characterization, causes of degradation, and food application. *Critical Reviews in Food Science and Nutrition, 60*(21). https://doi.org/10.1080/10408398.2019.1706444

Min, J. H., Kim, Y. H., Kim, J. H., Choi, S. Y., Lee, J. S., & Kim, H. K. (2012). Comparison of microbial diversity of Korean commercial makgeolli showing high β-glucan content and high antihypertensive activity, respectively. *Mycobiology, 40*(2), 138–141. https://doi.org/10.5941/MYCO.2012.40.2.138

Mirończuk-Chodakowska, I., Kujawowicz, K., & Witkowska, A. M. (2021). Beta-glucans from fungi: Biological and health-promoting potential in the covid-19 pandemic era. *Nutrients, 13*(11). https://doi.org/10.3390/nu13113960

Moriartey, S., Temelli, F., & Vasanthan, T. (2010). Effect of formulation and processing treatments on viscosity and solubility of extractable barley β-glucan in bread dough evaluated under in vitro conditions. *Cereal Chemistry, 87*(1), 65–72. https://doi.org/10.1094/CCHEM-87-1-0065

Muthuramalingam, K., Kim, Y., & Cho, M. (2021). β-glucan, "the knight of health sector": Critical insights on physiochemical heterogeneities, action mechanisms and health implications. *Critical Reviews in Food Science and Nutrition, 62*, 6908–6931. https://doi.org/10.1080/10408398.2021.1908221

Mykhalevych, A., Polishchuk, G., Nassar, K., Osmak, T., & Buniowska-Olejnik, M. (2022). β-glucan as a techno-functional ingredient in dairy and milk-based products—A review. *Molecules, 27*(19). https://doi.org/10.3390/molecules27196313

Nakashima, A., Yamada, K., Iwata, O., Sugimoto, R., Atsuji, K., Ogawa, T., Ishibashi-Ohgo, N., & Suzuki, K. (2018). B-Glucan in foods and its physiological functions. *Journal of Nutritional Science and Vitaminology, 64*(1), 8–17. https://doi.org/10.3177/jnsv.64.8

Rop, O., Mlcek, J., & Jurikova, T. (2009). Beta-glucans in higher fungi and their health effects. *Nutrition Reviews, 67*(11). https://doi.org/10.1111/j.1753-4887.2009.00230.x

Routray, W., & Orsat, V. (2012). Microwave-assisted extraction of flavonoids: A review. *Food and Bioprocess Technology, 5*(2). https://doi.org/10.1007/s11947-011-0573-z

Ruiz-Herrera, J., & Ortiz-Castellanos, L. (2019). Cell wall glucans of fungi. A review. *Cell Surface, 5*. https://doi.org/10.1016/j.tcsw.2019.100022

Schmidt, M. (2022). Cereal beta-glucans: An underutilized health endorsing food ingredient. *Critical Reviews in Food Science and Nutrition, 62*(12). https://doi.org/10.1080/10408398.2020.1864619

Schulze, M. B., Schulz, M., Heidemann, C., Schienkiewitz, A., Hoffmann, K., & Boeing, H. (2007). Fiber and magnesium intake and incidence of type 2 diabetes: A prospective study and meta-analysis. *Archives of Internal Medicine, 167*(9), 956. https://doi.org/10.1001/archinte.167.9.956

Shoukat, M., & Sorrentino, A. (2021). Cereal β-glucan: A promising prebiotic polysaccharide and its impact on the gut health. *International Journal of Food Science and Technology, 56*(5), 2088–2097. https://doi.org/10.1111/ijfs.14971

Suchecka, D., Gromadzka-Ostrowska, J., Żyła, E., Harasym, J. P., & Oczkowski, M. (2017). Selected physiological activities and health promoting properties of cereal beta-glucans. A review. *Journal of Animal and Feed Sciences, 26*(3). https://doi.org/10.22358/jafs/70066/2017

Synytsya, A., & Novák, M. (2013). Structural diversity of fungal glucans. *Carbohydrate Polymers, 92*(1). https://doi.org/10.1016/j.carbpol.2012.09.077

Tiwari, U., & Cummins, E. (2009). Factors influencing β-glucan levels and molecular weight in cereal-based products. *Cereal Chemistry, 86*(3). https://doi.org/10.1094/CCHEM-86-3-0290

Tosh, S. M., Brummer, Y., Wood, P. J., Wang, Q., & Weisz, J. (2004). Evaluation of structure in the formation of gels by structurally diverse (1 → 3)(1 → 4)-β-D-glucans from four cereal and one lichen species. *Carbohydrate Polymers, 57*(3). https://doi.org/10.1016/j.carbpol.2004.05.009

Utama, G. L., Dio, C., Lembong, E., Cahyana, Y., & Balia, R. L. (2020). Microorganism-based β-glucan production and their potential as antioxidant. *Systematic Reviews in Pharmacy, 11*(10). https://doi.org/10.31838/srp.2020.10.130

Vetvicka, V., Dvorak, B., Vetvickova, J., Richter, J., Krizan, J., Sima, P., & Yvin, J. C. (2007). Orally administered marine (1 → 3)-β-d-glucan Phycarine stimulates both humoral and cellular immunity. *International Journal of Biological Macromolecules, 40*(4), 291–298. https://doi.org/10.1016/j.ijbiomac.2006.08.009

Virkki, L., Johansson, L., Ylinen, M., Maunu, S., & Ekholm, P. (2005). Structural characterization of water-insoluble nonstarchy polysaccharides of oats and barley. *Carbohydrate Polymers, 59*(3), 357–366. https://doi.org/10.1016/j.carbpol.2004.10.006

Waszkiewicz-Robak, B. (2013). Spent brewer's yeast and beta-glucans isolated from them as diet components modifying blood lipid metabolism disturbed by an atherogenic diet. In *Lipid Metabolism*. InTech. https://doi.org/10.5772/51530

Watanabe, T., Shimada, R., Matsuyama, A., Yuasa, M., Sawamura, H., Yoshida, E., & Suzuki, K. (2013). Antitumor activity of the β-glucan paramylon from Euglena against preneoplastic colonic aberrant crypt foci in mice. *Food and Function, 4*(11), 1685–1690. https://doi.org/10.1039/c3fo60256g

Yiannikouris, A., François, J., Poughon, L., Dussap, C. G., Bertin, G., Jeminet, G., & Jouany, J. P. (2004). Alkali extraction of β-D-glucans from Saccharomyces cerevisiae cell wall and study of their adsorptive properties toward zearalenone. *Journal of Agricultural and Food Chemistry, 52*(11), 3666–3673. https://doi.org/10.1021/jf035127x

Yoshida, T., Honda, Y., Tsujimoto, T., Uyama, H., & Azuma, J. I. (2014). Selective isolation of β-glucan from corn pericarp hemicelluloses by affinity chromatography on cellulose column. *Carbohydrate Polymers, 111*, 538–542. https://doi.org/10.1016/j.carbpol.2014.04.050

Yuan, H., Lan, P., He, Y., Li, C., & Ma, X. (2020). Effect of the modifications on the physicochemical and biological properties of β-glucan-a critical review. *Molecules, 25*(1). https://doi.org/10.3390/molecules25010057

Zhang, C., Li, J., Wang, J., Song, X., Zhang, J., Wu, S., Hu, C., Gong, Z., & Jia, L. (2017). Antihyperlipidaemic and hepatoprotective activities of acidic and enzymatic hydrolysis exopolysaccharides from Pleurotus eryngii SI-04. *BMC Complementary and Alternative Medicine, 17*(1), 403. https://doi.org/10.1186/s12906-017-1892-z

Zhu, F., Du, B., & Xu, B. (2016). A critical review on production and industrial applications of beta-glucans. *Food Hydrocolloids, 52*, 275–288. https://doi.org/10.1016/j.foodhyd.2015.07.003

# Chapter 3
# Methods of Extraction of Beta-Glucan

**Shital Sargar, Swati Tiwari, G. Bhanuprakash Reddy,
and Digambar Kavitake**

## 3.1 Introduction

Beta-glucan, also denoted as β-glucan, is a type of biopolymer/polysaccharide of β-D-glucose present in a wide array of natural sources. These sources include cereal grains like wheat, rye, barley, and oats; they also include a variety of microorganisms including bacteria, fungi (yeast and edible mushrooms), algae, and others. Beta-glucans are composed of various glycosidic bonds, primarily β-(1, 6), β-(1, 4), and β-(1, 3) which connect glucose monomers (Maheshwari et al., 2017; Nie et al., 2018; Utama et al., 2021; Vezaro et al., 2022). Because of this, beta-glucans significantly differ in physicochemical and biological properties depending on a source from which they are extracted. Beta-glucans have been linked to a host for health benefits, including enhanced immune system performance, lowered cholesterol, weight management, control diabetes, and support for the prevention and treatment of many malignancies (Mantovani et al., 2008; Sonck et al., 2010; Tighe et al., 2010; Ludwig, 2000, 2003; Slavin, 2005; Haggard et al., 2013; de Araujo et al., 2017; Venn & Mann, 2004; De Munter et al., 2007). The food, cosmetic, medical, and pharmaceutical sectors use beta-glucans as functional additives because of their physico-functional and bioactive potentials (Havrlentová et al., 2011).

It is essential to get higher yield through methodical and efficient extraction from their sources to fully utilize beta-glucan in a variety of applications (Maheshwari et al., 2017). Numerous methods, each with advantages and disadvantages, have

S. Sargar
Department of Clinical Nutrition, MGM School of Biomedical Sciences, MGMIHS, Kamothe, Navi Mumbai, India

S. Tiwari
Department of Food Science and Technology, Pondicherry University, Pondicherry, India

G. B. Reddy · D. Kavitake (✉)
Biochemistry Division, ICMR- National Institute of Nutrition, Hyderabad, India



M. Bashir Mir, S. A. Mir (eds.), *Beta-Glucan: Sources, Properties and Applications*,
https://doi.org/10.1007/978-3-031-95788-8_3

been employed to extract beta-glucan from a range of sources. Various extraction techniques have been investigated and documented, contingent on the beta-glucans source (Maheshwari et al., 2017). These include methods of extraction assisted by ultrasound, microwaves, hot water, enzymatic, alkaline, and acid extraction. The type of plant or grain material, the required purity of the beta-glucans, and the equipment available all influence the method choice. This chapter has covered many commonly employed methods for obtaining beta-glucan from plants (such as cereal grains), fungi (such as yeasts and mushrooms), and bacteria. A comprehensive overview of the extraction techniques, including their advantages and limitations', effects on glucan quality, and variables influencing the yield and properties of beta-glucan is provided.

## 3.2 Extraction of Beta-Glucan

By examining the substantial research reported so far on techniques for both qualitative and quantitative extraction and the ensuing isolation of beta-glucan from natural sources, a comprehensive grasp of the selection of appropriate techniques and their alterations for further improvements in both quality and quantity can be achieved (Ahmad et al., 2009; Bhatty, 1995; Carr et al., 1990; Irakli et al., 2004; Papageorgiou et al., 2005; Skendi et al., 2003; Sasidharan et al., 2011; Zheng et al., 2011). The methods used to extract beta-glucan from these sources can be broadly classified into three major classes: enzymatic, physical, and chemical methods. The process of chemical extraction entails dissolving the source material's cell walls and releasing the beta-glucan by using chemicals such as alkali, acids, and solvents. In the acid extraction strategy (Ahluwalia & Ellis, 1984; Babu, 2015; Maheshwari et al., 2017), the source material is treated with an acid solution (citric acid) to extract the beta-glucan, or an alkaline solution (NaOH) in the alkaline extraction method. The hot water extraction approach, as described by Skendi et al. (2003), Irakli et al. (2004), and Papageorgiou et al. (2005), entails heating the source material to different temperatures and times to extract the beta-glucan. Proteases and amylases are specific enzymes that are used in the enzymatic extraction process to eliminate protein and starch impurities from the desired product. This method is commonly used because of its selectivity and ability to produce high-quality beta-glucan with minimal damage to the source material (Babu, 2015; Carr et al., 1990; Izydorczyk et al., 2000). Physical extraction methods that use physical forces to break down cell walls and extract beta-glucan include high-pressure homogenization, ultrasonication, and microfluidization (He et al., 2020; Izydorczyk et al., 2000; Tian et al., 2012; Zhu et al., 2016a; Liu et al., 2016). As can be seen in Fig. 3.1, there are several processes involved in the extraction, characterization, and application of beta-glucan. It is important to keep in mind that depending on the source material and the desired properties of the extracted beta-glucan, certain steps and specifications of the extraction process may vary.

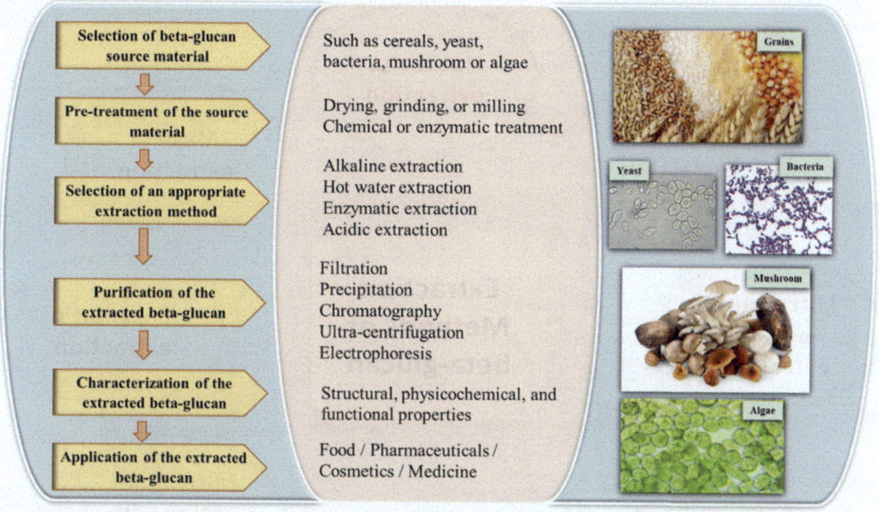


**Fig. 3.1** Comprehensive steps involved in the extraction process of beta-glucan

A sequence of milling and sifting procedures may be used in the extraction processes for dry beta-glucan to separate and concentrate the components. Using an abrasive cleaning method, the shell, pericarp, aleurone, sub-aleurone, and embryo layers are gradually removed until enriched fractions of the molecule are formed. Temperature, pH, type of solvent, length of extraction, and agitation all have an impact on the optimal conditions for beta-glucan extraction (Benito-Román et al., 2011). The extraction methods for dry beta-glucan sometimes involve a series of milling and sifting operations to separate and concentrate the substances. Sibakov et al. (2014) studied the impact of electrostatic separation and ultrafine grinding on enriched oat fractions to enhance the yield of beta-glucan with dry methods. Wet extraction methods for beta-glucan involve deactivating the natural enzymes present and extracting the compound using water or alkaline solutions. Hydrolytic enzymes or selective adsorption methods are used to remove impurities like starch and proteins. A solution containing 99% alcohol is used to precipitate beta-glucan (Mejía et al., 2020).

## 3.3 Extraction methods of Beta glucan

Over the past few years, a range of techniques has been employed to extract beta-glucan from various sources. As mentioned in Fig. 3.2, a few noteworthy extraction methods are hot water, enzymatic, acidic, alkaline, ultrasonic-assisted, microwave-assisted extraction, microfluidization, ultrasonic-assisted-hot water extraction, ultrasonic-microwave-assisted method, and pressurized liquid extraction method. These methods have been extensively documented for their effectiveness in beta-glucan extraction.

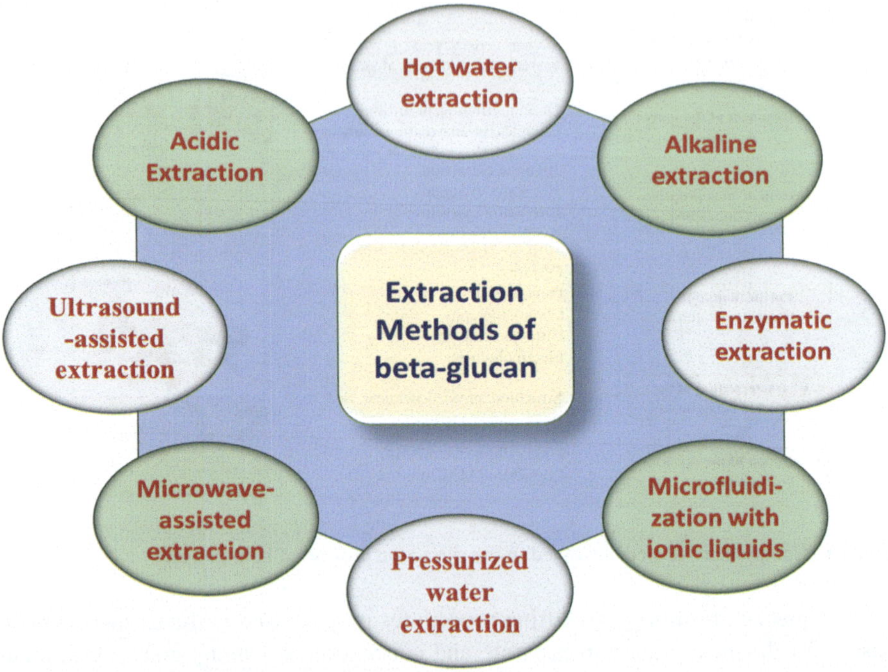


**Fig. 3.2** Extraction methods of beta-glucan

The procedure of treating cereal or plant material or samples with an alkaline solution, like sodium hydroxide (NaOH), effectively tears down the cell wall and releases the beta-glucans is known as the alkaline extraction method. The solution is then neutralized, and ethanol or isopropanol is used to precipitate the beta-glucans. The precipitate that results is then dried and collected by centrifugation (Maheshwari et al., 2017; Hu et al., 2015; Ahmad et al., 2009). Analogously, the sample material is treated with an acid solution (such as hydrochloric acid) in the acid extraction procedure. When the acid treatment breaks down the cell wall, the beta-glucans are liberated. The beta-glucans are precipitated using ethanol or isopropanol after neutralisation, and they are subsequently centrifuged and lyophilized (Ahmad et al., 2009; Maheshwari et al., 2017; Hu et al., 2015). As part of the hot water extraction process, the powdered sample is boiled in hot water to aid release the beta-glucans. When the solution is released, it is filtered to remove any leftover particles. Next, centrifugation, drying, and precipitation using ethanol or isopropanol (Ahmad et al., 2009; Maheshwari et al., 2017; Hu et al., 2015). In the enzymatic extraction process, particular enzymes are administered to the materials, such as cellulase, amylase, and protease. These enzymes effectively break down protein impurities, carbohydrates, and cell walls to aid in the release of beta-glucans. Following that, the solution is filtered to remove any remaining particles, and either ethanol or isopropanol is used to precipitate the beta-glucans (Daou & Zhang, 2012; Gamel et al., 2014; Hu et al., 2015; Maheshwari et al., 2017).

In the few stages of the ultrasonic-assisted extraction procedure, the sample or grain is mixed with the proper solvent and subjected to ultrasonic waves. Because of the cavitation caused by the ultrasonic vibrations, the solvent is subjected to significant pressure and temperature gradients. This facilitates the release of beta-glucan and promotes the breakdown of cell walls. After extraction, the fluid is typically filtered to remove any solid particles. Centrifugation can be utilized to collect the beta-glucans after ethanol or isopropanol precipitation, and they can then be dried (Zhang et al., 2023; Zhu et al., 2016b; Ying et al., 2011; Tian et al., 2012). For the microwave-assisted extraction process, the samples and a suitable solvent are mixed in a microwave-safe container. When the combination is exposed to microwave radiation, it warms up rapidly and boils in certain places. The breakdown of the cell walls and the release of beta-glucans into the solvent are aided by the heat generated. Ethanol or isopropanol precipitate precipitation is often carried out following the filtering of the resulting solution to remove any solid residues (Smiderle et al., 2017; Hu et al., 2019; Zhang et al., 2023). The microfluidization process is an additional means of extracting beta-glucan. Through the use of high-pressure homogenization, the cell walls of the plant material, powdered sample, or cell are broken down, releasing the beta-glucans. Samples are typically ground into a fine powder before the extraction process to increase the surface area for an efficient extraction. The powdered sample is mixed with a appropriate solvent or buffer solution to create a suspension. The suspension is then passed through a microfluidizer, a piece of high-pressure homogenization apparatus. The suspension travels through the microfluidizer's small chambers or channels under high pressure. The process breaks down the cell walls and releases the beta-glucans into the surrounding solution by producing strong shear and impact stresses. The resultant solution is usually filtered to get rid of any leftover solid particles after homogenization. After that, beta-glucans can be dried, collected by centrifugation, and precipitated using ethanol or isopropanol (Liu et al., 2016, 2020; Zhao et al., 2020). A thorough summary of the step-by-step variations that are part of the beta-glucan extraction process is given in Table 3.1. These variations include solvent type, the sample-to-solvent ratio, pH, temperature, incubation time, centrifugation parameters, and particular processing conditions like ultrasonic, microwave, or high pressure, etc.

## 3.4 Beta-Glucan Extraction from Plant Sources

Although beta-glucan can be found in many natural grain sources, according to Bacic et al. (2009) significant concentrations are only present in barley (2–20%), oat (3–8%), and sorghum (1–6.2%). In contrast, the quantities of beta-glucan in wheat and rye range from 0.5% to 1% and 1.4% to 2.6%, respectively (Henry, 1987; . Oat flakes (2.55–2.91%), pearl barley (2.95%), hulled barley (4.83%), and broken barley (3.96%) are a few examples of grain products and by-products that contain beta-glucan in varying amounts (Vizhi & Many, 2014). The extraction process may differ depending on the specific plant source and the desired characteristics of the extracted

**Table 3.1** An overview of major methods used for the extraction of beta-glucan from various sources

| Alkaline extraction | Acidic extraction | Hot water extraction | Enzymatic extraction | Ultrasonic- assisted extraction | Microwave- assisted extraction | Microfluidization |
|---|---|---|---|---|---|---|
| Sample / powder | Sample / powder | Sample / powder | Sample / powder | Sample / powder | Sample / powder | Sample / powder |
| Reflux with ethanol (80–85%) for 6 h | | | | Reflux with ethanol (80–85%) for 2–6 h | Add deionized water in 1:30 ratio (sample: water) | Add 200 mL of ionic solution EmimAc (1-ethyl-3-methylimidazolium acetate) in 1 g sample |
| Mix the sample with 1 M NaOH (in 1:7 ratio) | Mix the sample with 1 M Citric acid (in 1:7 Ratio) | Mix the sample with water (in 1:10 ratio) | Alpha amylase treatment for 3 h at 40 °C | Centrifugation for 15 min at 10,000 rpm | Subject in microwave extraction system with the power 850 W, frequency of 2455 MHz and pressure 30–35 bar | Heat at 85 °C and stirred for 30 min to 3 h |
| Allow mixing on magnetic stirrer at 55 °C (on hot plate) for 90 min | | | Centrifugation for 20 min at 15,000 rom and 40 °C | 80–160 W of ultrasonic power, 30–70 °C temperature, extraction time of 20–100 min, and water content of 5–25 mL per g of raw material sample | Centrifugation for 15 min at 10,000 rom | 30 min Dynamic High Pressure Microfluidization (DHPM) treatment at various pressure units (10,000–35,000 psi) |
| Centrifugation for 20 min at 15,000 and 40 °C | | | Protease enzyme was added to the supernatant and it was incubated for 3 h at 37 °C | Centrifuge at 10, 000 rpm for 15 min | Precipitation with two to three times the volume of 100% (v/v) ethanol and kept at 4 °C for the night | Precipitation with 2–3 times of volume of 100% (v/v) ethanol and kept at 4 °C for the night |

| | | | | | | |
|---|---|---|---|---|---|---|
| Supernatant taken & mixed with 1 M NaOH in 1:3 ratio | Supernatant taken & mixed with 1 M citric acid in 1:3 ratio | Supernatant adjusted at pH 8.5 with sodium bicarbonate and stirred on hot plate for 30 min at 55 °C | Centrifuged at 21,000 g for 25 min at 4 °C | Supernatant collected and deproteinized by Sevagmethod for 5 times | Centrifuge at 4500 rpm for 20 min | Centrifuge at 4500 rpm for 20 min |
| Centrifugation for 20 min at 18,000 g and 40 °C | | | Precipitation with 80% ethanol in equal volume with supernatant | 48–72 h dialysis against distilled water | Collect precipitates and lyophilize them to get dry sample | To remove the residualionic liquid, wash the precipitate with 4 times volume of absolute ethanol |
| With citric acid adjust the supernatant pH at 7 | With NaOH neutralize the supernatant pH at 7 | With citric acid adjust the supernatant pH at 4 | Centrifugation for 10–20 min at 4000 g | Lyophilization to get dry sample | – | Precipitates +600 ml of distilled water and evaporate the ethanol by heating at 65 °C for1 h with constant stirring |
| Centrifugation for 20–25 min at 21,000 g | | | Drying of pellets in a vacuum oven dryer or lyophilizer | – | – | Cool samples to room temperature under ambient conditions |
| Precipitation with 80% ethanol in equal volume with supernatant | | | – | – | – | Centrifuge at 4500 rpm for 30 min |
| Centrifugation for 10–20 min at 4000 g under cool environment (4 °C) | | | – | – | – | Collect the supernatant and lyophilize them to get dry sample |
| Dry pellets in vacuum oven or lyophilizer | | | – | – | – | |

**References:** Skendi et al., 2003; Papageorgiou et al., 2005; Irakli et al., 2004; He et al., 2020; Tian et al., 2012; Zhu et al., 2016a, b; Liu et al., 2016; Liu et al., 2020; Gil-Ramírez et al., 2019; Smiderle et al., 2017; Maheshwari et al., 2017

beta-glucan. Depending upon nature and type of beta-glucan (bond and branching nature) present within respective cereal, type of extraction method is selected. Solubility of beta-glucan in water or other solvent is another parameter which is taken in to consideration in selecting extraction method (Zhu & Wu, 2018). According to Sikora et al. (2013), beta-glucan is largely found in the endospore cell wall of oat, barley, wheat, and rice, or in the aleurone layer and sub-aleurone layer of cereals. There are two main techniques for extracting beta glucan from cereals: the dry extraction method and the wet extraction method. In the Dry extraction process, there is sequential removal of the aleurone shell, sub-aleurone pericarp embryo, using processes like milling, sieving, etc. This method requires a lengthy separation process, and the yield is often less than 20%. However, by using ultrafine grinding and an electrostatic separation process on oat fractions, the finished product had a 52.2% beta-glucan concentration (Sibakov et al., 2014; Benito-Román et al., 2013a). The beta-glucan is extracted through a wet extraction process using various solvents such as hot water, alkali, or acid. Using this technique, the natural enzyme beta-glucanase—which breaks down glucan—is deactivated. Additionally, contaminants like protein, lipid, and starch are removed using hydrolytic enzymes or a selective adsorption method (Izydorczyk & Dexter, 2008). The yield obtained through the wet extraction method is comparatively high than the dry extraction method; although the extraction yield is high, but there are certain limitation that comes along with this process like high viscosity of aqueous extract, the expensive method as it requires drying and concentration processes on large scale due to presence of large amount extraction liquid (Hu et al., 2015; Mejía et al., 2020; Yoo et al., 2020). A comprehensive analysis of the extraction and separation of beta-glucan is presented, focusing primarily on three grain sources: oat, barley, and wheat. This study investigates the effects of temperature, pH, and extraction medium on the yield and purity of beta-glucan that is produced using a variety of extraction techniques, such as enzymatic, acidic, and alkaline processes (Table 3.2). Based on the results, alkali and water extraction techniques are found to be more favorable due to their higher yield and effective recovery of beta-glucan (Maheshwari et al., 2017).

Following the use of several procedures, beta-glucan was extracted from barley and its yield, purity, and specific physicochemical and functional properties were assessed. The yield of beta-glucan in barley was reported using three distinct methods: alkaline extraction (5.6–11.9%; Bhatty, 1995), enzymatic extraction (5.24%; Ahmad et al., 2009), and acidic extraction (4–6%). Some observations on the yield and extraction of beta-glucan were reported by Ahmad et al. (2009). In contrast to alkaline extraction, which produced high yield and purity while assuring the most elimination of contaminants, he claimed that acidic extraction produced a lesser yield of beta-glucan than both enzymatic and hot water extraction method. In a subsequent work, molecular beta-glucan was extracted using a pressurised hot water extraction method by Benito-Román et al. (2013b). They concluded that utilizing higher temperatures and pressures reduces the extraction time. They determined that the optimal conditions for extraction were a temperature of 155 °C and a pressure of 50 bar, resulting in a high yield of 53.7%. However, it is important to note that beyond these values, the degradation of beta-glucan begins, leading to a

**Table 3.2** Extraction methods of beta-glucan from plant sources

| Source | Extraction method | Extraction solvents | Yield | References |
|---|---|---|---|---|
| Oat bran cereals | Alkali extraction | $NaOH/NaBH_4$ | 86% | Beer et al. (1997) |
| Barley grains and malt | Acidic extraction | Perchloric acid | 4.3–6% | Ahluwalia and Ellis (1984) |
| Barley bran | Alkaline extraction followed by neutralization with HCL | NaOH, HCL | ~49–79% | Bhatty (1995) |
| Oat bran | | NaOH, HCL | ~68–71% | Bhatty (1995) |
| Wheat bran | | NaOH, HCL | 23% | Cui et al. (2000) |
| Barley | Hot water extraction | $H_2O$ | 5.4% | Ahmad et al. (2009) |
| Barley | Enzymatic extraction | Enzymatic (α-amylase, protease), | 5.22% | Ahmad et al. (2009) |
| Barley | Acidic extraction | Acidic (Citric acid) | 4.65% | Ahmad et al. (2009) |
| Oats | Enzymatic extraction followed by ethanol precipitation | Protease, EtOH | 5.14% | Ahmad et al. (2010) |
| Red wheat bran | Ethanol and alkaline extraction | EtOH, $(NH_4)_2SO_4$ | 2.15% | Li et al. (2006) |
| White wheat bran | Ethanol and alkaline extraction | EtOH, $(NH_4)_2SO_4$ | 2.51% | Li et al. (2006) |
| Oat flour | Water extraction followed by ethanol precipitation | $H_2O$, EtOH | 85% | Skendi et al. (2003) |

decrease in its concentration. Making beta-glucan soluble in oat seeds is essential for its extraction. Assessing beta-glucan's solubility in water and alkali is a step in the extraction procedure. Following oat seed milling, isoelectric precipitation is used to extract beta-glucan from contaminants, primarily protein. In order to eliminate the contaminants and separate the beta-glucan, ethanol ammonium sulphate solution is added (Daou & Zhang, 2012). Research has demonstrated that temperature and pH levels during the process of extracting oat beta-glucan from a liquid solution are significant determinants for obtaining good recovery, in addition to the usual extraction approach (Dawkins & Nnanna, 1995). Additionally, a few investigations examined how well subcritical-water extraction (SWE) worked to separate beta-glucan from oat flour. Using high temperature and pressure settings (200 °C, pH 4.0 for the solvent, 10 min of extraction, and 425–850 μm for the particle size), SWE produced an extraction yield of 6.98 ± 1.17 g/100 g of oat flour for beta-glucan (Yoo et al., 2020). Approximately 2% of the total dietary fibre in whole wheat is composed of beta-glucan, which is comparatively low in wheat bran. Therefore, compared to oats and barley, the health advantages of consuming beta-glucan from wheat are probably limited. Cui et al. (2000, 2006) reported that an alkaline extraction method utilising sodium hydroxide was used to extract beta-glucan from wheat. The yield achieved from this process varied between 2.15 and 2.6 g/100 g of wheat. Depending on the plant species and extraction technique, beta-glucan derived from plant sources can have a variety of properties.

## 3.5 Beta-Glucan from Fungal Sources

Several fungal species, such as yeasts, mushrooms, and mycelia, contain the polysaccharide beta-glucan (Table 3.3). Over the past two decades, there has been a notable surge in the quantity of glucans generated by fungi, which has resulted in the identification of novel EPS molecular structures and the investigation of many strains, including the filamentous fungus *Lasiodiplodia theobromae* (Philippini et al., 2021). Glucans with the different structural patterns β (1 → 3, β (1 → 4), and β (1 → 6) play a crucial role in the utilization for various applications. Moreover, these glucans have also demonstrated positive impacts in the management of different diseases (Pelley & Strickland, 2000). The functional characteristics of the extracted beta-glucan can be impacted by variations in the molecular weight, solubility, and structure of the beta-glucan obtained from fungi. Because of its immunomodulatory, antitumor, and antioxidant properties, beta-glucan derived from fungi has potential uses in the food, cosmetics, and pharmaceutical industries (Mahmoud Amer et al., 2021; Chaiyasut et al., 2018; Chiozzi et al., 2021; Kusmiati & Dhewantara, 2016; Sánchez-Madrigal et al., 2015).

### *3.5.1 Mushroom*

When it comes to fungal sources specifically, the most widely available edible mushrooms on the global market are shiitake and oysters, which serve as the primary source of beta-glucans (Morales et al., 2021). *Pleurotus sajor-caju, Pleurotus eryngii,* and *Pleurotus ostreatoroseus* are edible mushrooms from which beta-glucan was isolated using hot water extraction procedures (Carbonero et al., 2006; Sakdasri et al., 2022). Beta-glucan extraction from dried *Pleurotus sajor-caju* mushroom (gray oyster) was extracted by pressurized hot water extraction method. The maximum β-glucan extraction yield (48.99 ± 1.57 g/100 g) was influenced by optimized parameters using central composite design, such as pressure (0.4–1.0 MPa), temperature (100–140 °C), and extraction time (20–60 min) (Sakdasri et al., 2022). Wu et al. (2013) employed the alkaline (sodium hydroxide) approach to extract beta-glucan from grey mushrooms (*Pleurotus tuber-regium sclerotia*), although the yield was lower (0.274 ± 0.044%) than with the

**Table 3.3** Extraction methods of beta-glucans from different fungal and bacterial sources

| Source/s | Microbial strains | Extraction method | Extraction solvents | Yield | Properties | References |
|---|---|---|---|---|---|---|
| ***Fungal sources*** | | | | | | |
| Mushroom (Shiitake variety) | *Lentinula edodes* | Ultrasound-assisted extraction | Water and Deep Eutectic Solvent (DES) | Water (23.47%) DES (25.11%) | Ferric reducing antioxidant power (FRAP) | Vezaro et al. (2022) |
| Mushroom | *Pleurouts Sajo-caju* | Microwave and Pressurized hot water extraction | Hot water (180 °C Pressure 0.4–1.0 MPa Time ~ 30 min) | 33–40% | NA | Smiderle et al. (2017) |
| | *Ganoderma lucidum* | Subcritical water extraction method | 163 °C,10 MPa time 130 min | 0.44 mg/100 g dry sample | NA | Askin et al. (2010) |
| | *Pleurotus tuber-regium sclerotia* | Alkali extraction (NaOH) | 100 °C 120 min | 0.274 ± 0.043 | NA | Wu et al. (2013) |
| | *Pleurotus eryngii* | Hot water extraction (temperature below 100 °C) Followed by ethanol precipitation | 100 °C, 360 min | 7.8 g/100 g | NA | Carbonero et al. (2006) |
| Commercial culture (Health Innovation Institute, Chiang Mai, Thailand) | *Saccharomyces cerevisiae* | Acid-alkaline method | Water | 3.72 g/L | Immuno-modulatory | Pengkumsri et al. (2016) |
| Commercial culture Fermipan® (Lesaffre Yeast Corporation, Wisconsin, USA) | *Saccharomyces cerevisiae* | Acid-alkaline method | Water | 3.94 mg | NA | Utama et al. (2021) |
| Commercial culture (Bioprocess Laboratory of the Faculty of Food Science and Nutritions Universiti Malaysia Sabah) | *Aspergillus oryzae* | Acid-alkaline method | Water | 82.5 mg | NA | Utama et al. (2021) |

(continued)

**Table 3.3** (continued)

| Source/s | Microbial strains | Extraction method | Extraction solvents | Yield | Properties | References |
|---|---|---|---|---|---|---|
| ***Bacterial sources*** | | | | | | |
| Commercial culture (Bioprocess Laboratory of the Faculty of Food Science and Nutritions Universiti Malaysia Sabah) | *Xanthomonas Campestris* | Acid-alkaline method | Water | 0.78 mg | NA | Utama et al. (2021) |
| | *Bacillus natto* | Acid-alkaline method | Water | 1.34 mg | NA | Utama et al. (2021) |
| Soil | *Rhizobium* punsense | Ethanol precipitation | Ethanol | 3.18 g/L | NA | Zhang et al. (2022) |
| Kefir grains | *Lactobacillus helveticus* | Ethanol precipitation | Ethanol | 124 mg/L | Prebiotic property | Wang et al. (2022) |
| Pickled cabbage | *Lactobacillus plantarus* H31 | Ethanol precipitation | Ethanol | NA | Upregulation of AMPK, AKT-2, and GLUT-4 expression in insulin resistant HepG2 cell line | Huang et al. (2020) |

*NA* not applicable

hot water method. A recent study employed a consecutive fractionation protocol involving cold water, hot water, and alkaline methods to extract β-glucans from three important mushroom species (*Pleurotus ostreatus Lentinula edodes,* and *Grifola frondosa*). *Pleurotus ostreatus* showed the highest concentration of β-glucan among these mushrooms, and it was concentrated in the more soluble portions. Conversely, *Grifola frondosa* possessed an elusive β-glucan that was predominantly found in the residue (7.38%), however, *Lentinula edodes* showed areas of β-glucans spread throughout both alkaline and aqueous extracts (Pérez-Bassart et al., 2023).

When it comes to shiitake mushrooms, ultrasound-assisted beta-glucan extraction proved to be successful when two solvents were utilised for the extraction process: water and NADE (Natural Deep Eutectic Solvent). NADE was discovered to be the most effective solvent between these two since it produced a higher yield of beta-glucan than water did when used as a solvent (Vezaro et al., 2022). In order to extract beta-glucan from A. bisporus, Tian et al. (2012) used the ultrasonic method under certain parameters. These comprised a 30 ml/g W/M ratio, a 70 °C extraction temperature, a 62-min extraction time, and a 230 W ultrasonic power. They were able to attain a maximum beta-glucan yield of 6.02% under these ideal circumstances. It is a complicated process that usually calls for a combination of at least two extraction techniques to extract beta-glucan from fungal sources. Among these methods include alkali, acid, hot water, microwave, ultrasonic, enzyme, and/or subcritical water extraction (Morales et al., 2019; Ruthes et al., 2016). For the objective of extracting beta-glucan from Ganoderma lucidum, Askin et al. (2010) combined hot water and pressure, which was found to be more effective than using only one parameter. In order to extract polysaccharides, specifically physiologically active β-glucan, from *Pleurotus ostreatus* and *Ganoderma lucidum* fruiting bodies, Smiderle et al. (2017) compared two highly advanced extraction methods: pressurized liquid extraction (PLE) and microwave-assisted extraction (MAE). The extraction of beta-glucans was successful using both techniques (33.6% and 40%, respectively). PLE offered the advantage of direct extract recovery without the need for centrifugation, whereas MAE required separating the extract from the residue. However, PLE had weaker reproducibility than MAE when tested with *P. ostreatus*, which resulted in a decrease in attractiveness. There were differences in the extracted polysaccharide concentrations, particularly in the glucans that were obtained, which may indicate differences in the selected mushroom species. While there are advantages and disadvantages to each technique, it has been demonstrated that MAE and PLE are easy, rapid, and efficient techniques to extract beta-glucans from mushrooms. Zhang et al. (2023) employed other techniques to isolate glucan from Dictyophora indusiate, namely ultrasonic-assisted extraction, microwave-assisted extraction, and ultrasonic-microwave-assisted extraction.

### 3.5.2 Yeast

Beta-glucans are important components found mostly in the innermost layer of yeast cell walls. Depending on the particular growth methods used, the amount of beta-glucan can vary significantly across different species of yeast (Javmen et al., 2012). Different glucan variations can be identified in the yeast cell wall based on their molecular branching patterns and bond types. According to Klis et al. (2002), beta-glucan makes up roughly 55–65% of the content of the cell wall in yeast (*S. cerevisiae*), which is a notable reservoir of the ingredient. Beta glucan extraction from yeast cells involves two processes. The initial stage of the beta-glucan extraction procedure is lysing yeast cells. This can be attained through various methods, including chemical ones such as using substances like NaOH, HCl, acetic acid, and citric acid. High pressure and sonication are two more physical methods (Wenger et al., 2008). Moreover, lysis of yeast cells has been achieved using enzymatic methods utilising lytic enzymes (Ferrer, 2006). At this point, the cytoplasm and the insoluble cell wall may be distinguished. The process's second step is removing the -glucan from the divided cell wall (Javmen et al., 2012). The most common techniques for removing beta-glucan from yeast are water extraction and acid-base extraction procedures. Sonication was used for 5 min in a pH eight phosphate buffer as part of the water extraction process to extract the contents of yeast cells. After autolysis, the solution was shaken well for 4 h at 30 °C, and then it was centrifuged for 15 min at 5000 × g. After two PBS washes, the yeast cells were treated with sodium dodecyl sulphate (SDS) to extract any possible protein materials. Eighty percent of the initial starter yeast cells were left after autoclaving for 2–6 h at a temperature between 80 and 121 °C. The water-insoluble gradient was subsequently subjected to centrifugation and drying (Fu et al., 2022; Mahmoud Amer et al., 2021). However, for the extraction of beta-glucan from yeast, the alkaline extraction method typically yields the highest output.

The acid-base extraction method is one of the methods that is more commonly used to extract beta-glucan from yeast sources. Using a yeast extract–glucose broth medium, the dry active yeast cells were grown for 48 h using this approach. Following the above-mentioned phase of collecting yeast cell pellets, the pellets were combined with 0.5–1 M NaOH, incubated for 2–6 h at 45–100 °C in a water bath, rewashed following centrifugation with distilled water, and finally, recentrifuged to acquire the concluding yield. The lysed yeast cell pellets were gathered and put to use in the extraction of beta-glucan. 0.5–2 M acetic acid (CH3COOH) was used for the extraction, and the sample was incubated for 1–2 h at 60–80 °C. The pellet was then collected by centrifugation at 6000 × g for 25 min at 4 °C, and it was then dried (Divya et al., 2020; Javmen et al., 2012; Pengkumsri et al., 2016; Indriati Ramadhani et al., 2020; Shokri et al., 2008). Additional findings showed that compared to the water extraction approach, a modified acid-base extraction method yields a greater biological efficacy of beta-glucan (Mahmoud Amer et al., 2021).

Additionally, the Dynamic High Pressure Microfluidization (DHPM) method was used to extract yeast beta-glucan by combining with ionic liquids (ILs) such as

AmimCl (1-allyl-3-methyl-imidazolium chloride), BmimAc (1-butyl-3-methyl-imidazolium acetate), and EmimAc (1-ethyl-3-methylimidazolium acetate). These ILs were used in high pressure microfluidization (HPM) to dissolve and alter the polysaccharide. EmimAc demonstrated good β-D-glucan dissolving properties, allowing for structural alterations for sustainable recycling. The ratio of -d-glucan to EmimAc, residence period, and DHPM pressure all had a significant impact on the solubility of -d-glucan. With a water solubility of 72.50% (w/w), the yield of water-soluble glucan was attained at 79.25% (w/w). Remarkably, the ionic liquids' structure remained intact after the DHPM treatment, enabling their recycling (Liu et al., 2016). The ionic liquid 1-butyl-3-methyl-imidazolium chloride (BmimCl) was utilised by Khanh et al. (2020) to extract beta-glucan from S. cerevisiae, a type of baker's yeast. Comparing the final beta-glucan to the amount of beta-glucan that was initially in the yeast, the yield was roughly 83.4% and the purity was 95.2%. It is yet unclear how high pressure microfluidization and ionic liquid together affect the physicochemical characteristics and structure of yeast beta-glucan. Moreover, additional research is necessary to determine the precise synergistic impact of these factors (Liu et al., 2016, 2020).

## 3.6 Beta Glucan Extraction from a Bacterial Source

Bacterial origin beta-glucans (β-glucans) are polysaccharides often referred as exo-polysaccharides (EPS), generally produced by bacteria mainly in response to environmental stresses. In general, EPS are high molecular weight and long chain polysaccharides which are made up of either one or different types of monosaccharides. So, they are categorized as homo- or hetero - polysaccharides based on the composition of repeating units. Glucan EPS are mainly categorized under HoPS with curdlan as the important example of β-glucan while α-glucan are of four types including dextrans, alternans, mutans, and reuterans. HePS has a more complex synthesis pathway, whereas HoPS has a more straightforward one. Aside from the extracellular enzymes' production, no active transit phases or energy consumption are involved. These extracellular enzymes function as the glycosyl donor during the production of HoPS and are known as glycosyltransferases or glucansucrase (Ryan et al., 2015).

The variations in structure, size, and composition make them versatile and useful natural additives for the food and pharma industry by virtue of their ability to improve techno-functional properties as well as their associated health benefits. Various bacteria are found to produce β-glucan EPS including *Lactobacillus fermentum, Lactobacillus suebicus, Pediococcus parvulus, Xanthomonas campertis,* etc. (Utama et al. 2021). The major advantage of glucan EPS derived from bacterial sources is its generally recognized safe (GRAS) status. This paves the way for *in situ* as well as *ex situ* EPS enrichment as food additives (Zannini et al., 2016). Hence, owning to their hydrocolloid-like properties, glucans or glucan EPS-producing strains can be directed to various industrial applications. Unlike, cereal-based β-glucans which mainly work on cholesterol reduction with potential benefits on cardiovascular health; bacterial glucans act on modifying the immune system and maintaining healthy functioning of cells. They are known to alleviate a substantial number of clinical disorders including their anti-cancer, anti-inflammatory, and anti-infective properties as the promising ones (Murphy et al., 2020). The other biological roles of β-glucans include their prebiotic, immunomodulatory, and antioxidant activity ; Wang et al., 2022).

This emphasizes that production and extraction under controlled conditions are the need of the hour on account of their numerous technological, biological, and economic advantages. Moreover, these microbial products have the potential to replace their synthetic counterparts as well. The extraction of β-glucan may vary depending upon their sources, for instance among microorganisms the extraction method of β-glucan for bacteria is entirely different from the extraction method of yeast as described in previous sections. Since β-glucan is a kind of extracellular polymeric substance or exopolysaccharide i.e. EPS. Hence, the commonly used extraction method for β-glucan follows the normal EPS extraction protocol. In general, the fermented broth is centrifuged at high rpm (usually between 8000 to 13, 000) under refrigerated conditions to separate the cells. To this cell-free supernatant, chilled ethanol is added and kept overnight at 4 °C to precipitate the EPS. This is further centrifuged to collect the EPS and then washed using distilled water (Yadav et al., 2011). However, EPS extracted from fermentation media is crude EPS may contain some impurities such as proteins, salts, etc. Hence, dialysis is one of the most commonly used techniques to obtain partially purified EPS (Jayamanohar et al., 2018; Zhang et al., 2022). It is noteworthy, that further purification including the removal of nucleic acid impurities (such as bacterial DNA and RNA) of partially purified EPS can be done using column chromatographic techniques. Size exclusion chromatography is commonly used to obtain pure EPS fractions which facilitates the determination of their molecular weight as well. This is further freeze-dried to obtain the final EPS powder.

Recently, membrane filtration has come into light for the withdrawal of polysaccharides from bacterial sources which is based upon molecular weight cut-off principle. These techniques include ultrafiltration, microfiltration, and diafiltration those can be carried out independently or in combinations with ethanol precipitation to increase the effectiveness (Levander et al., 2001). For *Lactobacillus plantarum* and *Lactococcus lactis*, ultrafiltration was found to be more effective than ethanol precipitation (Ziadi et al., 2018). Typical examples of various membranes used are polyethersulfone and regenerated cellulose (Zhang et al., 1998). Ultrasonic and microwave-assisted extractions are two more cutting-edge methods that have been employed to extract these polysaccharides (Chen et al., 2012; Zeng et al., 2012). These are more effective at producing larger yields of polysaccharides, but they are only effective with fungi; perhaps because of this, they haven't been applied to bacteria. As a result, similar non-thermal methods may be used in future studies to extract EPS from bacterial sources.

## 3.7 Factors Affecting Yield and Purity of Beta Glucan Extraction

Various factors impact the yield and purity of beta-glucan extracted, as illustrated in Fig. 3.3. First and foremost, the extraction technique selected is crucial. Different techniques, such as acidic extraction, hot water extraction, enzymatic extraction, alkaline extraction, and other aided extraction methods, might yield varying quantities of beta-glucan. The availability of different grains, such as wheat, barley, or oats, has an impact on the extraction yield as well. Furthermore, the extraction conditions—which include elements like temperature, pH, and extraction time can have a significant impact on the production of beta-glucan (Gangopadhyay et al., 2015). Additional factors that may impact the yield of extracted beta-glucan include the extraction solvent's concentration, the solvent-to-sample ratio, particle size, stirring rate, and the existence of impurities or interfering substances (Benito-Román et al., 2011). It is important to comprehend and enhance these variables to attain elevated beta-glucan extraction yields. When purifying beta-glucan from diverse sources, a variety of techniques are used, each having pros and cons related to the purity and characteristics of the beta-glucan. To acquire the most effective purity of the beta-glucan, impurities such as cell wall components, lipids, proteins, and insoluble complex polysaccharides should be removed. The sample was heated under reflux at various times while suspended in petroleum ether for the defatting treatment. After centrifugation, the leftovers underwent various acetone washes. Following the deproteinization process, the material was combined with a mixture of composite

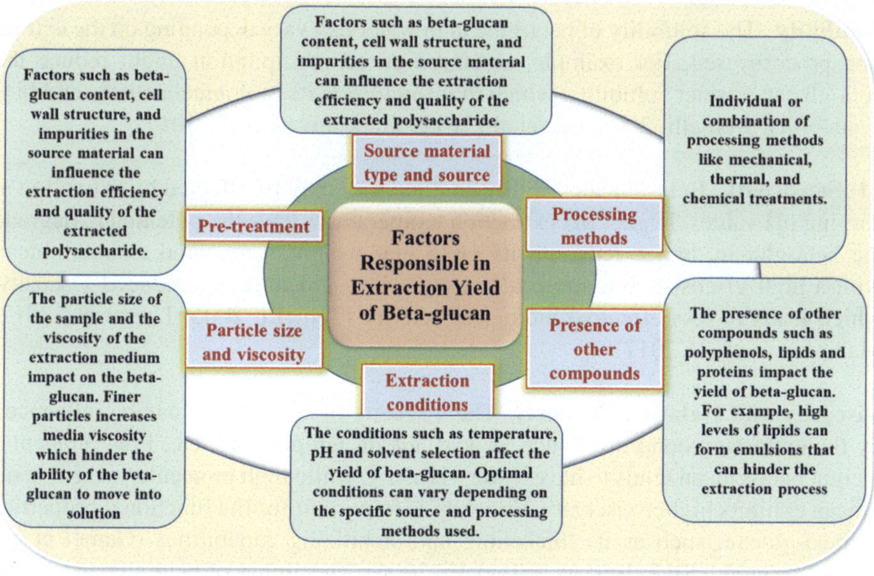


**Fig. 3.3** Factors responsible for extraction yield of beta-glucan

enzymes, primarily proteases (papain, bromelain, and neutral protease), and subjected to varying periods of reaction at 50 °C. As a result, β-glucan samples containing a range of protein levels were produced (Szwengiel & Stachowiak, 2016; Fu et al., 2022). Precipitation of the beta-glucan also influences the extraction yield. Solvent ethanol should be used in a sufficient quantity to precipitate the complete beta-glucan extracted from the source. Insufficient ethanol leads to a lower extraction yield of beta-glucan. Dehydration or drying is one of the techniques which affects the properties of the extracted beta-glucan. Three distinct procedures were used to dry the β-glucan samples: freeze-drying, hot air drying, and vacuum drying.

## 3.8 Effect of Extraction Methods on Properties of Beta-Glucan

The physicochemical and functional properties of beta-glucan can be significantly influenced by the extraction techniques used to get it. The qualities of the finished product are greatly influenced by several extraction-related variables, including pH, temperature, extraction duration, centrifugal gravity, purification, and isolation techniques (Maheshwari et al., 2017).

**Temperature Sensitivity** High temperatures during the extraction process can produce beta-glucan with a larger molecular weight. On the other hand, beta-glucan may degrade at temperatures exceeding 160 °C, which would lower its molecular weight (Temelli, 2006; Lee et al., 2012; Maheshwari et al., 2017).

**Solubility** The solubility of beta-glucan in water can vary depending on the extraction process used. For example, alcohol-induced precipitation might reduce the beta-glucan's water solubility, which in turn impacts its biological activity and bioavailability (Temelli, 2006; Lee et al., 2012; Maheshwari et al., 2017).

**pH Sensitivity** Beta-glucan stability and integrity may be affected by extraction at varying pH values. Higher pH extraction temperatures have the potential to degrade the beta-glucan, hence reducing its molecular weight. Compared to beta-glucan with a high viscosity, low molecular weight beta-glucan has a reduced viscosity, which leads to less effective biological activity (Temelli, 2006; Lee et al., 2012; Maheshwari et al., 2017).

**Viscosity and Gelation Property** The viscosity of beta-glucan can be influenced by the extraction conditions such as acid, alkali, temperature, etc. Low molecular weight beta-glucan tends to have lower viscosity, while high molecular weight beta-glucan exhibits higher viscosity. Viscosity is important for the functional properties of beta-glucan, such as its thickening and stabilizing capabilities (Gamel et al., 2014; Lee et al., 2012; Burkus & Temelli, 2006). The ability of beta-glucan to form gels is affected by the extraction method, and some methods may result in beta-glucan with poor gelling properties. The effects of oat MW and β-glucan content on gel characteristics were examined. Thus, for various purposes, the addition of sugar or modification of the molecular weight fraction ratios might alter the textural characteristics and melting profiles of β-glucan gels (Brummer et al., 2014; Aktas-Akyildiz et al., 2018). A significant variation in extraction methods was observed to water binding capacity and foaming capacity of beta-glucan from barley. Alkaline and acidic extraction methods adversely affect the viscosity properties. A significant effect of extraction methods was also observed on the color parameters of β-glucan (Ahmad et al., 2009).

The purification treatment to remove fat increased oil-binding ability and a decrease in the swelling power of beta-glucan. Utilizing ball milling for disruption further enhanced the oil-binding capacity while reducing the water-holding capacity. In contrast, freeze-drying exhibited superior water-holding capacity, oil-binding capacity, and swelling power when compared to vacuum drying and hot air drying (Fu et al., 2022).

**Biological Activity** The extraction process can impact the biological activity of beta-glucan. Factors like molecular weight, solubility, and viscosity influence its effectiveness in various biological actions, including immune modulation, cholesterol-lowering effects, and antitumor activities (Yuan et al., 2019; Mishra et al., 2020). Also processes like deproteinization, defatting treatment, ball milling (except ball milling), and freeze-drying significantly improved the bile acid binding capacity and showed increased inhibitory effects against pancreatic lipase, α-amylase, and α-glucosidase. These enhancements play a crucial role in the in vitro hypolipidemic and hypoglycaemic properties of β-glucan (Fu et al., 2022). β-glucan extracted by the acid-base approach demonstrated a promising antibacterial, antitoxic, immunostimulatory, and anticancer action (Mahmoud Amer et al., 2021; Mishra et al., 2020; Yuan et al., 2019). The choice of extraction method and the specific conditions employed can greatly influence the physicochemical and functional properties of beta-glucan, making it essential to carefully optimize the extraction process to achieve desired characteristics for specific applications.

## 3.9 Advantages and Limitations of Beta-Glucan Extraction Methods

### *3.9.1 Aqueous Extraction*

***Advantages*** Simplicity and cost-effectiveness. Scalability, making it suitable for large-scale production. Minimal chemical usage, resulting in a cleaner extract. Preservation of beta-glucan's biological activity.

***Limitations*** Susceptibility to degradation at high temperatures. Potential extraction of undesirable compounds along with beta-glucan. Limited extraction efficiency, requiring optimization for maximum yield.

### *3.9.2 Alkaline Extraction*

***Advantages*** Enhanced extraction efficiency compared to aqueous extraction. Effective removal of impurities, resulting in a purer beta-glucan extract. Preservation of beta-glucan's biological activity.

***Limitations*** Potential degradation of beta-glucan at high pH levels. Requirement for careful control of pH and temperature to prevent degradation. Generation of alkaline waste, necessitating proper disposal.

### *3.9.3 Enzymatic Extraction*

***Advantages*** Highly specific and selective extraction of beta-glucan. Preservation of beta-glucan's structure and biological activity. Mild reaction conditions, minimizing degradation risks. Environmentally friendly, as it does not involve harsh chemicals.

***Limitations*** Requirement for enzyme optimization to achieve maximum extraction efficiency. Potential for enzyme contamination of the beta-glucan extract. Relatively high cost of enzymes, limiting its feasibility for large-scale production.

### *3.9.4 Microwave-Assisted Extraction*

***Advantages*** Rapid and efficient extraction process, reducing extraction time significantly. Enhanced extraction efficiency due to improved mass transfer and penetration. Preservation of beta-glucan's biological activity due to reduced exposure to heat. Potential for selective extraction of specific beta-glucan fractions.

***Limitations*** Requirement for specialized equipment and expertise. Potential for localized overheating and degradation of beta-glucan. Careful optimization of extraction parameters to minimize degradation risks.

### 3.9.5 Ultrasound-Assisted Extraction

***Advantages*** Enhanced extraction efficiency due to cavitation-induced disruption of cell walls. Reduced extraction time compared to conventional methods. Preservation of beta-glucan's biological activity due to milder extraction conditions. Potential for selective extraction of specific beta-glucan fractions.

***Limitations*** Requirement for specialized equipment and expertise. Potential for localized overheating and degradation of beta-glucan. Careful optimization of extraction parameters to minimize degradation risks.

## 3.10 Conclusion

Beta-glucan, a type of polysaccharide, has garnered considerable attention among researchers and professionals due to its wide-ranging health benefits, including its ability to modulate the immune system, reduce cholesterol levels, and mitigate the risk of chronic diseases. As a result, there is a growing demand for beta-glucan extraction methods that efficiently isolate this valuable compound from various sources, such as cereals, fungi, and algae. Ultimately, the choice of extraction method depends on factors like the source material, desired quality, and intended application of the beta-glucan extract. Furthermore, the high-order structure of these polysaccharides is still unknown, and the connection between their structure and biological activity is not well understood because of the structural diversity of the polysaccharide molecules, including that of their conformation and complex chains. Consequently, more research into the precise structure and biological activities is required.

The chapter delves into a thorough examination of the many extraction techniques used to extract beta-glucans from different sources, emphasizing the benefits and drawbacks of each method. It explores the key elements influencing the extraction procedure and looks closely at the characteristics of the resultant beta-glucan. Several factors must be considered when choosing an extraction technique, such as the type of source material, the desired level of purity for the beta-glucan extract, and the intended use of the extracted material.

This knowledge gap emphasizes the need for further in-depth investigation into the precise structure of beta-glucans and their corresponding biological activities. Addressing this gap is crucial for progressing an understanding of these molecules and unlocking their full potential for various applications. Therefore, continued research is necessary to elucidate the intricate details of beta-glucan structures and establish a clearer link between their structural configurations and biological functions.

## References


Ahluwalia, B., & Ellis, E. E. (1984). A rapid and simple method for the determination of starch and β-glucan in barley and malt. *Journal of the Institute of Brewing, 90*(4), 254–259.

Ahmad, A., Anjum, F. M., Zahoor, T., Nawaz, H., & Din, A. (2009). Physicochemical and functional properties of barley β-glucan as affected by different extraction procedures. *International Journal of Food Science & Technology, 44*(1), 181–187.

Ahmad, A., Anjum, F. M., Zahoor, T., Nawaz, H., & Ahmed, Z. (2010). Extraction and characterization of β-D-glucan from oat for industrial utilization. *International Journal of Biological Macromolecules, 46*(3), 304–309.

Ahmad, A., Munir, B., & Abrar, M. (2012). Perspective of β-glucan as functional ingredient for food industry. *Journal of Nutrition and Food Sciences, 02*(2), 2. https://doi.org/10.4172/2155-9600.1000133

Aktas-Akyildiz, E., Sibakov, J., Nappa, M., Hytönen, E., Koksel, H. A. M. İ. T., & Poutanen, K. (2018). Extraction of soluble β-glucan from oat and barley fractions: Process efficiency and dispersion stability. *Journal of Cereal Science, 81*, 60–68.

Askin, R., Sasaki, M., & Goto, M. (2010). Recovery of water-soluble compounds from *Ganoderma lucidum* by hydrothermal treatment. *Food and Bioproducts Processing, 88*(2–3), 291–297.

Babu, L. (2015). Green extraction techniques, structural analysis and antioxidant activities of β-glucan present in oats. *International Journal of Latest Trends in Engineering and Technology, 5*, 125–135.

Bacic, A., Fincher, G. B., & Stone, B. (Eds.). (2009). *Chemistry, biochemistry, and biology of 1–3 beta glucans and related polysaccharides*. Academic.

Beer, M. U., Wood, P. J., & Weisz, J. (1997). Molecular weight distribution and (1→ 3)(1→ 4)-β-D-glucan content of consecutive extracts of various oat and barley cultivars. *Cereal Chemistry, 74*(4), 476–480.

Benito-Román, O., Alonso, E., & Lucas, S. (2011). Optimization of the β-glucan extraction conditions from different waxy barley cultivars. *Journal of Cereal Science, 53*(3), 271–276.

Benito-Román, Ó., Alonso, E., & Cocero, M. J. (2013a). Ultrasound-assisted extraction of β-glucans from barley. *LWT – Food Science and Technology, 50*(1), 57–63. https://doi.org/10.1016/j.lwt.2012.07.006

Benito-Román, Ó., Alonso, E., & Cocero, M. J. (2013b). Pressurized hot water extraction of β-glucans from waxy barley. *Journal of Supercritical Fluids, 73*, 120–125. https://doi.org/10.1016/j.supflu.2012.09.014

Bhatty, R. S. (1995). Laboratory and pilot plant extraction and purification of β-glucans from hull-less barley and oat brans. *Journal of Cereal Science, 22*(2), 163–170. https://doi.org/10.1016/0733-5210(95)90046-2

Brummer, Y., Defelice, C., Wu, Y., Kwong, M., Wood, P. J., & Tosh, S. M. (2014). Textural and rheological properties of oat beta-glucan gels with varying molecular weight composition. *Journal of Agricultural and Food Chemistry, 62*(14), 3160–3167.

Carbonero, E. R., Gracher, A. H. P., Smiderle, F. R., Rosado, F. R., Sassaki, G. L., Gorin, P. A. J., & Iacomini, M. (2006). A β-glucan from the fruit bodies of edible mushrooms *Pleurotus eryngii* and *Pleurotus ostreatoroseus*. *Carbohydrate Polymers, 66*(2), 252–257. https://doi.org/10.1016/j.carbpol.2006.03.009

Carr, J. M., Glatter, S., Jeraci, J. L., & Lewis, B. A. (1990). Enzymic determination of β-glucan in cereal-based food products. *Cereal Chemistry, 67*(3), 226–229.

Chaiyasut, C., Pengkumsri, N., Sivamaruthi, B. S., Sirilun, S., Kesika, P., Saelee, M., et al. (2018). Extraction of β-glucan of *Hericium erinaceus*, *Avena sativa* L., and *Saccharomyces cerevisiae* and in vivo evaluation of their immunomodulatory effects. *Food Science and Technology, 38*, 138–146.

Chen, W., Wang, W. P., Zhang, H. S., & Huang, Q. (2012). Optimization of ultrasonic-assisted extraction of water-soluble polysaccharides from *Boletus edulis* mycelia using response surface methodology. *Carbohydrate Polymers, 87*(1), 614–619.

Chiozzi, V., Eliopoulos, C., Markou, G., Arapoglou, D., Agriopoulou, S., El Enshasy, H. A., & Varzakas, T. (2021). Biotechnological addition of β-glucans from cereals, mushrooms and yeasts in foods and animal feed. *Processes, 9*(11), 1889.

Cui, W., Wood, P. J., Blackwell, B., & Nikiforuk, J. J. C. P. (2000). Physicochemical properties and structural characterization by two-dimensional NMR spectroscopy of wheat β-D-glucan—Comparison with other cereal β-D-glucans. *Carbohydrate Polymers, 41*(3), 249–258.

Daou, C., & Zhang, H. (2012). Oat beta-glucan: Its role in health promotion and prevention of diseases. *Comprehensive Reviews in Food Science and Food Safety, 11*(4), 355–365.

Dawkins, N. L., & Nnanna, I. A. (1995). Studies on oat gum [(1→ 3, 1→ 4)-β-D-glucan]: composition, molecular weight estimation and rheological properties. *Food Hydrocolloids, 9*(1), 1–7.

De Araújo, T. V., Andrade, E. F., Lobato, R. V., Orlando, D. R., Gomes, N. F., De Sousa, R. V., et al. (2017). Effects of beta-glucans ingestion (*Saccharomyces cerevisiae*) on metabolism of rats receiving high-fat diet. *Journal of Animal Physiology and Animal Nutrition, 101*(2), 349–358.

De Munter, J. S. L., Hu, F. B., Spiegelman, D., Franz, M., & Van Dam, R. M. (2007). Whole grain, bran, and germ intake and risk of type 2 diabetes: A prospective cohort study and systematic review. *PLoS Medicine, 4*(8), e261.

Divya, M., Gopi, N., Iswarya, A., Govindarajan, M., Alharbi, N. S., Kadaikunnan, S., ... & Vaseeharan, B. (2020). β-glucan extracted from eukaryotic single-celled microorganism Saccharomyces cerevisiae: Dietary supplementation and enhanced ammonia stress tolerance on Oreochromis mossambicus. *Microbial Pathogenesis, 139*, 103917.

Kusmiati, & Dhewantara, F. X. (2016). Cholesterol-lowering effect of beta glucan extracted from *Saccharomyces cerevisiae* in rats. *Scientia Pharmaceutica, 84*(1), 153–165.

Ferrer, P. (2006). Revisiting the *Cellulosimicrobium cellulans* yeast-lytic β-1, 3-glucanases toolbox: A review. *Microbial Cell Factories, 5*(1), 10. https://doi.org/10.1186/1475-2859-5-10

Fu, W., Zhao, G., & Liu, J. (2022). Effect of preparation methods on physiochemical and functional properties of yeast β-glucan. *LWT-Food Science and Technology, 160*, 113284.

Gamel, T. H., Abdel-Aal, E. S. M., Ames, N. P., Duss, R., & Tosh, S. M. (2014). Enzymatic extraction of beta-glucan from oat bran cereals and oat crackers and optimization of viscosity measurement. *Journal of Cereal Science, 59*(1), 33–40.

Gangopadhyay, N., Hossain, M. B., Rai, D. K., & Brunton, N. P. (2015). Optimisation of yield and molecular weight of β-glucan from barley flour using response surface methodology. *Journal of Cereal Science, 62*, 38–44.

Gil-Ramírez, A., Smiderle, F. R., Morales, D., Iacomini, M., & Soler-Rivas, C. (2019). Strengths and weaknesses of the aniline-blue method used to test mushroom (1→ 3)-β-d-glucans obtained by microwave-assisted extractions. *Carbohydrate Polymers, 217*, 135–143.

Haggård, L., Andersson, M., & Punga, A. R. (2013). β-glucans reduce LDL cholesterol in patients with myasthenia gravis. *European Journal of Clinical Nutrition, 67*(2), 226–227.

Havrlentova, M., Petrulakova, Z., Burgarova, A., Gago, F., Hlinkova, A., & Šturdík, E. (2011). β-glucans and their significance for the preparation of functional foods-a review. *Czech Journal of Food Sciences, 29*(1), 1–14.

He, J. L., Guo, H., Wei, S. Y., Zhou, J., Xiang, P. Y., Liu, L., et al. (2020). Effects of different extraction methods on the structural properties and bioactivities of polysaccharides extracted from Qingke (Tibetan hulless barley). *Journal of Cereal Science, 92*, 102906.

Henry, R. J. (1987). Pentosan and (1→ 3),(1→ 4)-β-glucan concentrations in endosperm and wholegrain of wheat, barley, oats and rye. *Journal of Cereal Science, 6*(3), 253–258. https://doi.org/10.1016/S0733-5210(87)80062-0

Hu, X., Zhao, J., Zhao, Q., & Zheng, J. (2015). Structure and characteristic of β-glucan in cereal: A review. *Journal of Food Processing and Preservation, 39*(6), 3145–3153.

Hu, W., Zhao, Y., Yang, Y., Zhang, H., Ding, C., Hu, C., et al. (2019). Microwave-assisted extraction, physicochemical characterization and bioactivity of polysaccharides from *Camptotheca acuminata* fruits. *International Journal of Biological Macromolecules, 133*, 127–136.

Huang, Z., Lin, F., Zhu, X., Zhang, C., Jiang, M., & Lu, Z. (2020). An exopolysaccharide from *Lactobacillus plantarum* H31 in pickled cabbage inhibits pancreas α-amylase and regulating metabolic markers in HepG2 cells by AMPK/PI3K/Akt pathway. *International Journal of Biological Macromolecules, 143*, 775–784.

Indriati Ramadhani, D. L., Yuliani, Y., & Amir, M. (2020). Extraction, characterization, and biological toxicity of β-glucans from Saccharomyces cerevisiae isolated from ragi. *Journal of Microbial Systematics and Biotechnology, 2*(2), 35–43.

Irakli, M., Biliaderis, C. G., Izydorczyk, M. S., & Papadoyannis, I. N. (2004). Isolation, structural features and rheological properties of water-extractable β-glucans from different Greek barley cultivars. *Journal of the Science of Food and Agriculture, 84*(10), 1170–1178.

Izydorczyk, M. S., & Dexter, J. E. (2008). Barley β-glucans and arabinoxylans: Molecular structure, physicochemical properties, and uses in food products–a review. *Food Research International, 41*(9), 850–868. https://doi.org/10.1016/j.foodres.2008.04.001

Izydorczyk, M. S., Storsley, J., Labossiere, D., MacGregor, A. W., & Rossnagel, B. G. (2000). Variation in total and soluble β-glucan content in hulless barley: Effects of thermal, physical, and enzymic treatments. *Journal of Agricultural and Food Chemistry, 48*(4), 982–989.

Javmen, A., Grigiškis, S., & Gliebutė, R. (2012). β-glucan extraction from *Saccharomyces cerevisiae* yeast using *Actinomyces rutgersensis* 88 yeast lyzing enzymatic complex. *Biologija, 58*, 51–59.

Jayamanohar, J., Devi, P. B., Kavitake, D., Rajendran, S., Priyadarisini, V. B., & Shetty, P. H. (2018). Characterization of α-D-glucan produced by a probiont *Enterococcus hirae* KX577639 from feces of south Indian Irula tribals. *International Journal of Biological Macromolecules, 118*, 1667–1675.

Klis, F. M., Mol, P., Hellingwerf, K., & Brul, S. (2002). Dynamics of cell wall structure in *Saccharomyces cerevisiae*. *FEMS Microbiology Reviews, 26*(3), 239–256. https://doi.org/10.1111/j.1574-6976.2002.tb00613.x

Lee, S. H., Jang, G. Y., Kim, K. J., Lee, M. J., Kim, T. J., Lee, J., & Jeong, H. S. (2012). Effect of temperature, solvent concentration, and pH on the β-glucan extraction. *The Korean Journal of Food And Nutrition, 25*(4), 871–877.

Levander, F., Svensson, M., & Rådström, P. (2001). Small-scale analysis of exopolysaccharides from *Streptococcus thermophilus* grown in a semi-defined medium. *BMC Microbiology, 1*(1), 1–5.

Li, W., Cui, S. W., & Kakuda, Y. (2006). Extraction, fractionation, structural and physical characterization of wheat β-D-glucans. *Carbohydrate Polymers, 63*(3), 408–416.

Liu, H., Li, Y., Gao, J., Shi, A., Liu, L., Hu, H., et al. (2016). Effects of microfluidization with ionic liquids on the solubilization and structure of β-d-glucan. *International Journal of Biological Macromolecules, 84*, 394–401.

Liu, H., Bai, W., He, L., Li, X., Shah, F., & Wang, Q. (2020). Degradation mechanism of *Saccharomyces cerevisiae* β-D-glucan by ionic liquid and dynamic high-pressure microfluidization. *Carbohydrate Polymers, 241*, 116123.

Ludwig, D. S. (2000). Dietary glycemic index and obesity. *The Journal of Nutrition, 130*(2), 280S–283S.

Ludwig, D. S. (2003). Dietary glycemic index and the regulation of body weight. *Lipids, 38*(2), 117–121.

Maheshwari, G., Sowrirajan, S., & Joseph, B. (2017). Extraction and isolation of β-glucan from grain sources—A review. *Journal of Food Science, 82*(7), 1535–1545.

Mahmoud Amer, E., Saber, S. H., Abo Markeb, A., Elkhawaga, A. A., Mekhemer, I. M. A., Zohri, A. A., Abujamel, T. S., Harakeh, S., & Abd-Allah, E. A. (2021). Enhancement of β-glucan biological activity using a modified acid–base extraction method from *Saccharomyces cerevisiae*. *Molecules, 26*(8), 2113. https://doi.org/10.3390/molecules26082113

Mantovani, M. S., Bellini, M. F., Angeli, J. P. F., Oliveira, R. J., Silva, A. F., & Ribeiro, L. R. (2008). β-Glucans in promoting health: Prevention against mutation and cancer. *Mutation Research/Reviews in Mutation Research, 658*(3), 154–161.

Mejía, S. M. V., de Francisco, A., & Bohrer, B. (2020). A comprehensive review on cereal β-glucan: Extraction, characterization, causes of degradation, and food application. *Critical Reviews in Food Science and Nutrition, 60*(21), 3693–3704.

Mishra, N., Mishra, N., & Mishra, P. (2020). Influence of different extraction methods on physiochemical and biological properties of β-glucan from indian barley varieties. *Carpathian Journal of Food Science & Technology, 12*(1).

Morales, D., Smiderle, F. R., Villalva, M., Abreu, H., Rico, C., Santoyo, S., ... & Soler-Rivas, C. (2019). Testing the effect of combining innovative extraction technologies on the biological activities of obtained β-glucan-enriched fractions from Lentinula edodes. *Journal of Functional Foods, 60*, 103446.

Morales, D., Shetty, S. A., López-Plaza, B., Gómez-Candela, C., Smidt, H., Marín, F. R., & Soler-Rivas, C. (2021). Modulation of human intestinal microbiota in a clinical trial by consumption of a β-D-glucan-enriched extract obtained from Lentinula edodes. *European Journal of Nutrition, 60*, 3249–3265.

Murphy, E. J., Rezoagli, E., Major, I., Rowan, N. J., & Laffey, J. G. (2020). β-glucan metabolic and immunomodulatory properties and potential for clinical application. *Journal of Fungi, 6*(4), 356.

Nilsson, M., Åman, P., Härkönen, H., Hallmans, G., Knudsen, K. E. B., Mazur, W., & Adlercreutz, H. (1997). Content of nutrients and lignans in roller milled fractions of rye. *Journal of the Science of Food and Agriculture, 73(2), 143–148. https://doi.org/10.1002/(SICI)1097-0010(199702)73:2<143::AID-JSFA698>3.0.CO;2-H*

Nie, S., Cui, S. W., & Xie, M. (2018). Chapter 9—Cereal beta-glucan. In S. Nie, S. W. Cui, & M. Xie (Eds.), *Bioactive Polysaccharides, Academic Press* (pp. 445–482). https://doi.org/10.1016/B978-0-12-809418-1.00003-4.

Papageorgiou, M., Lakhdara, N., Lazaridou, A., Biliaderis, C. G., & Izydorczyk, M. S. (2005). Water extractable (1→ 3, 1→ 4)-β-D-glucans from barley and oats: An intervarietal study on their structural features and rheological behaviour. *Journal of CerealScience, 42(2), 213–224.*

Pelley, R. P., & Strickland, F. M. (2000). Plants, polysaccharides, and the treatment and prevention of neoplasia. *Critical Reviews in Oncogenesis, 11(3&4), 189–225.*

Pengkumsri, N., Sivamaruthi, B. S., Sirilun, S., Peerajan, S., Kesika, P., Chaiyasut, K., & Chaiyasut, C. (2016). Extraction of β-glucan from Saccharomyces cerevisiae: Comparison of different extraction methods and in vivo assessment of immunomodulatory effect in mice. *Food Science and Technology, 37, 124–130.*

Pérez-Bassart, Z., Fabra, M. J., Martínez-Abad, A., & López-Rubio, A. (2023). Compositional differences of β-glucan-rich extracts from three relevant mushrooms obtained through a sequential extraction protocol. *Food Chemistry, 402, 134207.*

Philippini, R. R., Martiniano, S. E., Franco Marcelino, P. R., Chandel, A. K., Dos Santos, J. C., & Da Silva, S. S. (2021). Production of β-glucan exopolysaccharide lasiodiplodan by Lasiodiplodia theobromae CCT 3966 from corn bran acid hydrolysate. *Applied Microbiology and Biotechnology, 105, 2319–2332*

Ruthes, A. C., Smiderle, F. R., & Iacomini, M. (2016). Mushroom heteropolysaccharides: A review on their sources, structure and biological effects. *Carbohydrate Polymers, 136*, 358–375. https://doi.org/10.1016/j.carbpol.2015.08.061

Ryan, P. M., Ross, R. P., Fitzgerald, G. F., Caplice, N. M., & Stanton, C. (2015). Sugar-coated: Exopolysaccharide producing lactic acid bacteria for food and human health applications. *Food and Function, 6*(3), 679–693.

Sakdasri, W., Arnutpongchai, P., Phonsavat, S., Bumrungthaichaichan, E., & Sawangkeaw, R. (2022). Pressurized hot water extraction of crude polysaccharides, β-glucan, and phenolic compounds from dried gray oyster mushroom. *LWT- Food Science and Technology, 168*, 113895. https://doi.org/10.1016/j.lwt.2022.113895

Sánchez-Madrigal, M. Á., Neder-Suárez, D., Quintero-Ramos, A., Ruiz-Gutiérrez, M. G., Meléndez-Pizarro, C. O., Piñón-Castillo, H. A., Galicia-García, T., & Ramírez-Wong, B. (2015). Physicochemical properties of frozen tortillas from nixtamalized maize flours enriched with β-glucans. *Food Science and Technology, 35*(3), 552–560. https://doi.org/10.1590/1678-457X.6715

Sasidharan, S., Chen, Y., Saravanan, D., Sundram, K. M., & Latha, L. Y. (2011). Extraction, isolation and characterization of bioactive compounds from plants' extracts. *African journal of traditional, complementary and alternative medicines, 8*(1).

Shokri, H., Asadi, F., & Khosravi, A. R. (2008). Isolation of β-glucan from the cell wall of Saccharomyces cerevisiae. *Natural Product Research, 22*(5), 414–421.

Sibakov, J., Abecassis, J., Barron, C., & Poutanen, K. (2014). Electrostatic separation combined with ultra-fine grinding to produce β-glucan enriched ingredients from oat bran. *Innovative Food Science and Emerging Technologies, 26*, 445–455. https://doi.org/10.1016/j.ifset.2014.10.004

Sikora, P., Tosh, S. M., Brummer, Y., & Olsson, O. (2013). Identification of high β-glucan oat lines and localization and chemical characterization of their seed kernel β-glucans. *Food Chemistry, 137*(1–4), 83–91. https://doi.org/10.1016/j.foodchem.2012.10.007

Skendi, A., Biliaderis, C. G., Lazaridou, A., & Izydorczyk, M. S. (2003). Structure and rheological properties of water soluble β-glucans from oat cultivars of *Avena sativa* and *Avena bysantina*. *Journal of Cereal Science, 38*(1), 15–31.

Slavin, J. L. (2005). Dietary fiber and body weight. *Nutrition, 21*(3), 411–418.

Smiderle, F. R., Morales, D., Gil-Ramírez, A., de Jesus, L. I., Gilbert-López, B., Iacomini, M., & Soler-Rivas, C. (2017). Evaluation of microwave-assisted and pressurized liquid extractions to obtain β-d-glucans from mushrooms. *Carbohydrate Polymers, 156*, 165–174.

Sonck, E., Stuyven, E., Goddeeris, B., & Cox, E. (2010). The effect of β-glucans on porcine leukocytes. *Veterinary Immunology and Immunopathology, 135*(3–4), 199–207.

Szwengiel, A., & Stachowiak, B. (2016). Deproteinization of water-soluble ß-glucan during acid extraction from fruiting bodies of *Pleurotus ostreatus* mushrooms. *Carbohydrate Polymers, 146*, 310–319.

Temelli, F. (2006). Extraction and functional properties of barley β-glucan as affected by temperature and pH. *Journal of Food Science, 62*(6), 1194–1201.

Tian, Y., Zeng, H., Xu, Z., Zheng, B., Lin, Y., Gan, C., & Lo, Y. M. (2012). Ultrasonic-assisted extraction and antioxidant activity of polysaccharides recovered from white button mushroom (*Agaricus bisporus*). *Carbohydrate Polymers, 88*(2), 522–529.

Tighe, P., Duthie, G., Vaughan, N., Brittenden, J., Simpson, W. G., Duthie, S., et al. (2010). Effect of increased consumption of whole-grain foods on blood pressure and other cardiovascular risk markers in healthy middle-aged persons: A randomized controlled trial. *The American Journal of Clinical Nutrition, 92*(4), 733–740.

Utama, G. L., Dio, C., Sulistiyo, J., Chye, F. Y., Lembong, E., Cahyana, Y., et al. (2021). Evaluating comparative β-glucan production aptitude of *Saccharomyces cerevisiae*, *Aspergillus oryzae*, *Xanthomonas campestris*, and *Bacillus natto*. *Saudi Journal of Biological Sciences, 28*(12), 6765–6773.

Venn, B. J., & Mann, J. I. (2004). Cereal grains, legumes and diabetes. *European Journal of Clinical Nutrition, 58*(11), 1443–1461.

Vezaro, F. D., da Rosa, B. V., & Kuhn, R. C. (2022). Ultrasound-assisted extraction of β-glucans from *Lentinula edodes* using natural deep eutectic solvent and water. *Journal of Chemical Technology and Biotechnology, 97*(12), 3306–3316. https://doi.org/10.1002/jctb.7190

Vizhi, V. K., & Many, J. N. (2014). Study on estimation, extraction and analysis of barley beta-glucan. *International Journal of Science and Research, 3*, 1480–1484.

Wang, X., Xu, M., Xu, D., Ma, K., Zhang, C., Wang, G., et al. (2022). Structural and prebiotic activity analysis of the polysaccharide produced by *Lactobacillus helveticus* SNA12. *Carbohydrate Polymers, 296*, 119971.

Wenger, M. D., DePhillips, P., & Bracewell, D. G. (2008). A microscale yeast cell disruption technique for integrated process development strategies. *Biotechnology Progress, 24*(3), 606–614. https://doi.org/10.1021/bp070359s

Wood, P. J., Siddiqui, I. R., & Paton, D. (1978). Extraction of high-viscosity gums from oats. *Cereal Chemistry, 55*, 1038–1049.

Wu, G. H., Hu, T., Huang, Z. L., & Jiang, J. G. (2013). Characterization of water and alkali-soluble polysaccharides from *Pleurotus tuber-regium sclerotia*. *Carbohydrate Polymers, 96*(1), 284–290. https://doi.org/10.1016/j.carbpol.2013.03.036

Yadav, V., Prappulla, S. G., Jha, A., & Poonia, A. (2011). A novel exopolysaccharide from probiotic *Lactobacillus fermentum* CFR 2195: Production, purification and characterization. *Biotechnology, Bioinformatics and Bioengineering, 1*(4), 415–421.

Ying, Z., Han, X., & Li, J. (2011). Ultrasound-assisted extraction of polysaccharides from mulberry leaves. *Food Chemistry, 127*(3), 1273–1279.

Yoo, H. U., Ko, M. J., & Chung, M. S. (2020). Hydrolysis of beta-glucan in oat flour during subcritical-water extraction. *Food Chemistry, 308*, 125670. https://doi.org/10.1016/j.foodchem.2019.125670

Yuan, H., Lan, P., He, Y., Li, C., & Ma, X. (2019). Effect of the modifications on the physicochemical and biological properties of β-glucan—A critical review. *Molecules, 25*(1), 57.

Zannini, E., Waters, D. M., Coffey, A., & Arendt, E. K. (2016). Production, properties, and industrial food application of lactic acid bacteria-derived exopolysaccharides. *Applied Microbiology and Biotechnology, 100*, 1121–1135.

Zeng, W. C., Zhang, Z., Gao, H., Jia, L. R., & Chen, W. Y. (2012). Characterization of antioxidant polysaccharides from *Auricularia auricular* using microwave-assisted extraction. *Carbohydrate Polymers, 89*(2), 694–700.

Zhang, L., Zhou, J., Yang, G., & Chen, J. (1998). Preparative fractionation of polysaccharides by columns packed with regenerated cellulose gels. *Journal of Chromatography A, 816*(2), 131–136.

Zhang, B., Zhang, Z., Lyu, X., & Zhao, W. (2022). Structural characterization of a water-soluble and antimicrobial β-glucan secreted by *Rhizobium pusense. Journal of Bioscience and Bioengineering, 134*(4), 318–325.

Zhang, Y., Lei, Y., Qi, S., Fan, M., Zheng, S., Huang, Q., & Lu, X. (2023). Ultrasonic-microwave-assisted extraction for enhancing antioxidant activity of *Dictyophora indusiata* polysaccharides: The difference mechanisms between single and combined assisted extraction. *Ultrasonics Sonochemistry, 95*, 106356.

Zhao, Y., Zhou, H. M., Huang, Z. H., & Zhao, R. Y. (2020). Different aggregation states of barley β-glucan molecules affect their solution behavior: A comparative analysis. *Food Hydrocolloids, 101*, 105543.

Zheng, X., Li, L., & Wang, Q. (2011). Distribution and molecular characterization of β-glucans from hull-less barley bran, shorts and flour. *International Journal of Molecular Sciences, 12*(3), 1563–1574.

Zhu, F., Du, B., & Xu, B. (2016a). A critical review on production and industrial applications of beta-glucans. *Food Hydrocolloids, 52*, 275–288.

Zhu, W., Xue, X., & Zhang, Z. (2016b). Ultrasonic-assisted extraction, structure and antitumor activity of polysaccharide from Polygonum multiflorum. *International Journal of Biological Macromolecules, 91*, 132–142.

Zhu, M., & Wu, S. (2018). The growth performance and nonspecific immunity of loach Paramisgurnus dabryanus as affected by dietary β-1, 3-glucan. *Fish & shellfish immunology, 83*, 368–372

Ziadi, M., Bouzaiene, T., M'Hir, S., Zaafouri, K., Mokhtar, F., Hamdi, M., & Boisset-Helbert, C. (2018). Evaluation of the efficiency of ethanol precipitation and ultrafiltration on the purification and characteristics of exopolysaccharides produced by three lactic acid bacteria. *BioMed Research International, 2018.*

# Chapter 4
# Chemistry of Beta-Glucan

**Zahida Naseem, Sajad Ahmad Mir, Painuly Sudhanshu, and J. Sudhakara Rao**

## 4.1 Introduction

Beta-glucan (*β*-glucan) comprise a group of β-D-glucose naturally occurring in the cell walls of cereals (such as oat, barley, rye etc), bacteria and fungus, with significantly differing properties depending on source. Beta-glucan is a soluble fiber that has been gaining interest due to its multiple functional and bioactive properties. It plays a beneficial role in insulin resistance, dyslipidemia, hypertension and obesity (Nishantha et al., 2018). The linkage patterns of *β*-glucans consist of *β*-(1,3)- or *β*-(1,4)-glucan backbone with either (1,2)- or (1,6)-linked *β*-glucopyranosyl side branches. It is suggested that yeast derived beta-(1,3; 1,6)-glucan is more biologically active than its counterparts with predominantly 1,3; 1,4 linkages. In industries like food and pharmaceuticals, carbohydrates are crucial sources of energy. The consumption of carbohydrates from plants, including polymers that are soluble and insoluble fiber, is necessary for both human and animal diets. The behavior of dietary fibers in water is typically classified as either insoluble or soluble and they differ in terms of their properties. The insoluble fibers, such as cellulose, lignin and some hemicelluloses are not disintegrated when water is added. In contrast, when dissolved in water, soluble fibers like hemicelluloses and pectin form thick gel-like solutions. A thick layer of water can be formed in the gastrointestinal tract by soluble fibers, which has the ability to inhibit the absorption of sugars and lipids. Their ability to reduce postprandial glucose spikes makes them particularly beneficial in treating conditions like diabetes. Similar physiological effects have been observed in foods that contain beta-glucans, particularly those found in high amounts, such as

Z. Naseem
Division of Food Science and Technology, SKUAST-Kashmir, Shalimar, Srinagar, India

S. A. Mir (✉) · P. Sudhanshu · J. Sudhakara Rao
Department of Life Science (Microbiology and Food Science and Technology), GITAM (Deemed to be University), Visakhapatnam, Andhra Pradesh, India

 
M. Bashir Mir, S. A. Mir (eds.), *Beta-Glucan: Sources, Properties and Applications*,
https://doi.org/10.1007/978-3-031-95788-8_4

oats and barley, beans and certain gums (Wang et al., 2017). Polymers such as β-glucan are believed to be responsible for the observed biological effects, given their high molecular weight and tendency to occur in alcohol precipitate fractions.

## 4.2 Sources of β-Glucan

The amount of β-glucan that can be obtained from natural sources varies depending on the source of the β-glucan, which defines its chemical structure, functional and technological features, as well as growing circumstances (for grain crops), technology, method and degree of purification (Demirbas, 2005). Beta-glucan is one of the important constituents of cell walls of most fungi and many plants. The major sources of β-glucan are cereal crops: oats, barley, rye, wheat, etc., which also justifies the vast information available on these raw materials. The beta-glucan content differs largely among the various sources. Among the cereals, greater amount of beta-glucan is found in barley followed by oat while as least amount is found in wheat and rice. Beta-glucan found in cereal brans like barley and oat is usually produced as agricultural by-products. Depending on the type of plant, grain crops can yield up to 6% of β-glucan. Utilizing genotyping technology can enhance the production of β-glucan. This technique has been used to grain crops like wheat, barley, and oats, to yield raw materials with an average β-glucan content of 1–2% for wheat and 6–9% for oats and barley. Further, increases in the content of β-glucan can be achieved by distinct, unique genotyping techniques (Zhang et al., 2021). The percentage of beta-glucan in oats ranges between 4.5% and 5.5% and it is mainly present in the aleuronic and sub aleuronic layers. In mushroom, the beta-glucan content varied between 3.1% and 46.5% and mainly concentrated in mycelia and/or fruiting bodies. In yeast, it ranges between 5% and 7% and is found mainly in the outer cell wall of yeasts like *Saccharomyces cerevisiae,Candida milleri, Kloeckera apiculate, Debaryomyces hansenii* and *Zygosaccharomyces bailii*. Bacteria are also the potential source of beta-glucan; a linear curdlan structure is produced by *Agrobacterium, Bacillus* and *Pseudomonas* spp. during fermentation (Fig. 4.1, Table 4.1).

## 4.3 Structure and Biochemistry

A non-starch polysaccharide, beta-glucan is made up of linear D-glucose chains with a (1–3)-β-D-glycosidic linkage, and side chains that are connected to β-(1–6) in fungi and β-(1–4) in plants and bacteria. Oats include a random unbranched linear structure of β-(1–3) and β-(1–4)-D-glycosidic linkage. It contains 70% of β-(1–4) and 30% of β-(1–3)-glycosidic bonds in its structure. The presence of β-(1–3) linkage breaks the uniform structure of oat beta-glucan, allowing it to form viscous solution upon solubilisation (Xu et al., 2013). The structure of wheat

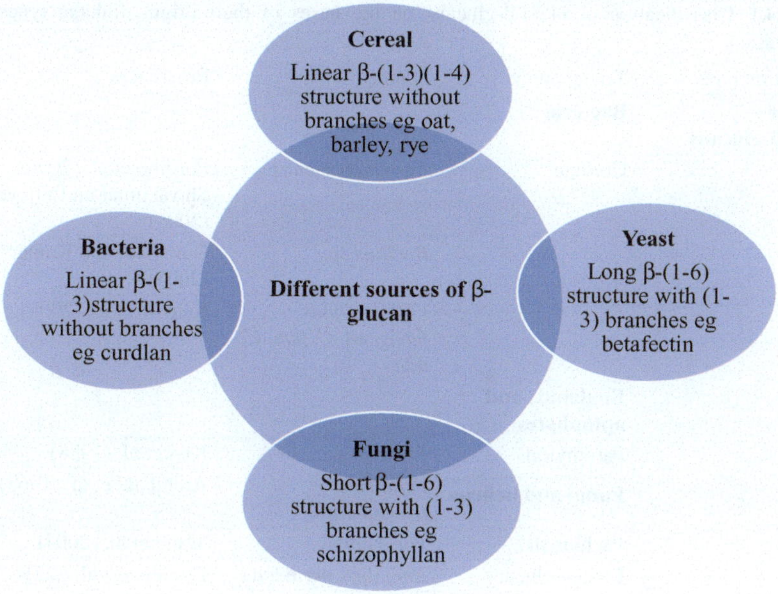


**Fig. 4.1** Sources of beta-glucan

beta-glucan is quite regular due to the predominance of trimeric unit that contributes to its high gelling ability. Beta-glucan derived from yeast and mushroom composed of β-(1–3) linkage with occasional β-(1–6) linkages, has greater biological activity, than cereal beta-glucan (β-(1–3) and β-(1–4)). Fungal beta-glucan is a mixture of linear β-(1–3) linkage with short branch chains connected through β-(1–6) (Nakashima et al., 2018) while as bacterial beta-glucan has linear unbranched chain of β-(1–3)-D-glucan backbone. Seaweed/algae beta-glucan is comprised of straight chain of β-(1–3)-D-glucan with high levels of β-(1–6) branches. Beta- glucan is divided into two classes, soluble having β-(1–3, 1–6)-D-linked glucose and insoluble having of β-(1–3, 1–4)-D-linked glucose. Based on source, beta-glucan can exist in a range of conformations such as random coils, helices (single, double, or triple), rod like shape and wormlike shapes.

Cyclic glucans structure and size distribution are subject to variation, which is determined by the type of material they come from and the processing procedures employed. The formation of cyclic configurations in carbohydrates, known as β-glycosidic bonds, is achieved by interconnecting glucose units in a group of amino acids called cytoamyloses or glycoglucans, which are also commonly referred to as polysaccharides. The most common type of cyclic glucans is a group called cytodextrins, which are made from starch through enzymatic action and are known for their specialized properties, such as reducing dietary fibers. The majority of cyclic assemblies in cyclodextrins are made up of six, seven, or eight glucose units, with the former being a ring-shaped structure. Cyclic glucans typically contain different amounts of glucose units, leading to the formation of rings with

**Table 4.1** Classification of (1,3)-β-glucans on the bases of their origin, linkage types, and organization

| Structural type | Trivial name | Source | References |
|---|---|---|---|
| **Linear (1,3)-β-glucans** | **Bacteria** | | |
| | Curdlan | *Agrobacterium* sp. | Portilho et al. (2006), Shivakumar and Vijayendra (2006) |
| | | *Bacillus* sp. | Gummadi and Kumar (2005) |
| | | *Cellulomonas flavigena, C. fimi, C. uda* | Kenyon et al. (2005) |
| | **Euglenids and aptophytes** | | |
| | Paramylon | *Euglena gracilis* | Kiss et al. (1988) |
| | **Fungi and lichens** | *Peranema trichophorum* | Archibald et al. (1963) |
| | Pachyman | *Poria cocos* | Wang et al. (2004) |
| | Lichen glucans | *Ramalina peruviana* | Cordeiro et al. (2004) |
| | | *Cladonia* spp., *Umbilicaria mammulata* | Carbonero et al. (2002) |
| **Linear (1,3)-β-glucans with (1,6)-linked-β-glucosylorβ-(1,6)-oligoglucosyl side chains** | **Chromists Phaeophytes** | | |
| | Laminarin | *Laminaria* spp. | Chizhov et al. (1998) |
| | **Oomycetes** | | |
| | Mycolaminarin | *Phytophthora* sp. | Wang and Bartnicki-Garcia (1980) |
| | | *Pythium apinadermatum* | Blaschek et al. (1992) |
| | | *Achyla bisexualis* | Lee et al. (1996) |
| | | *Phytophthora parasitica* | Bruneteau et al. (1988) |
| | **Chrysophytes** | | |
| | Chrysolaminarin | *Ochromonas malhamensis* | Archibald et al. (1963) |
| | **Fungi** | | |
| | Epiglucan | *Flammulina velutipes* | Smiderle et al. (2006) |
| | | *Ganoderma lucidum* | Chang and Lu (2004) |
| | Lentinan | *Lentinus edodes* | Zhang et al. (2004) |
| | | *Pleurotus florida* | Rout et al. (2005) |
| | Schizophyllan | *Schizophyllum commune* | Akima et al. (1985) |

(continued)

**Table 4.1** (continued)

| Structural type | Trivial name | Source | References |
|---|---|---|---|
| **Branch-on-branch (1,3;1,6)- β-glucans** | Fungal wall glucan | *Aspergillus fumigatus* | Fontaine et al. (2000) |
| | Oomycete wall glucan | *Pythium aphnidermatum* | Blaschek et al. (1992) |
| | Yeast wall glucan | *Saccharomyces cerevisiae* | Kollar et al. (1997) |
| | | *Candida albicans* | Ruiz-Herrera et al. (2006) |
| **Linear (1,3;1,4)- β-Glucans** | Cereal glucans | *Hordeum vulgare, Avena sativa, Triticum vulgare* | Cui et al. (2000), Lazaridou et al. (2004), Papageorgiou et al. (2005) |
| | Horsetail glucan | *Equisetum arvense* | Fry et al. (2008) |

varying sizes, from small to larger. The functional properties of cyclic glucans are highly diverse, that is why they are frequently used in food industry, pharmaceuticals and materials engineering. Nuclear magnetic resonance (NMR) spectroscopy, mass spectrometry (MS) and chromatography are commonly employed to analyze the structure and distribution of ring sizes in cyclic glucans. These analytical methods provide valuable information about the molecular structure, conformational organization and physicochemical properties of cyclic (and other) glucans so that they may be better characterized and used in industrial applications as well as in scientific research (Fig. 4.2).

## *4.3.1 Structural Analysis*

### 4.3.1.1 Physicochemistry

The physicochemical traits of (1,3)-β-glucans are primarily due to their inherent stability and the distinctive ordered conformations observed in hydrated conditions. Linear (1,3)-β-glucans exhibit a triple helix in anhydrous and hydrated solid states. The presence of side-chain-substituted (1,3)-β-glucans having similar diffraction patterns implies that they share the same triple helical structure, but with less lateral aggregation of guanidine backbones due to integration of side chains (Gidley & Nishinari, 2009).

On the other hand, there is limited experimental evidence for ordered conformations in (1,3; 1,4)-β-glucans, but it is believed that well-defined repeating backbone sequences could aid in facilitating inter-chain interactions. The chemical composition, molecular size, biological origin and thermal and solvent treatment histories of polysaccharides play a role in determining the characteristics of 1,3- β-glucans as crystalline gel molecules. Although these variables have comparatively simple primary structures, their interaction leads to a wide range of behaviours. These properties include not only the macro-structural organization in solutions, gels or cellular matrices but also the underlying ‘cell-mediated’ interactions associated with promising therapeutic properties of (1,3)-β-glucans.

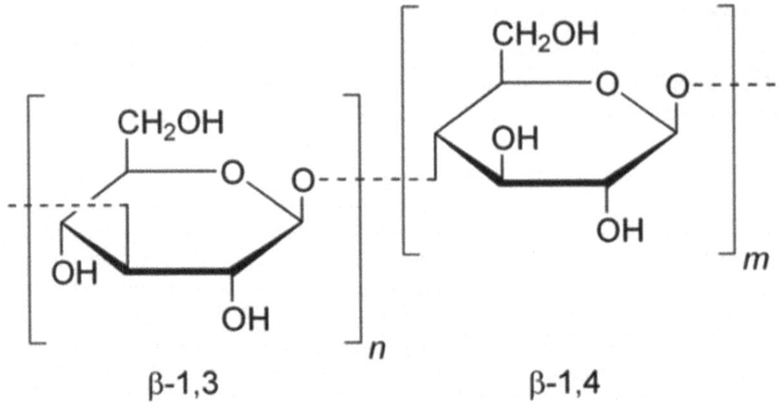


**Fig. 4.2** Structure of beta glucan

#### 4.3.1.2 Chain Stiffness

Cyclic monomeric units also known as monosaccharide's are typically connected by glycosidic bonds in polysaccharides. Due to the cyclic nature of monomer units and its inherent rigidity, these structures are highly resistant to polysaccharide chains. Unlike most other rigid linkages, (1,6)-linkages have three-bond glycosidic connections, which allows for greater conformational flexibility. When they are dissolved, the stiff chains cause high viscosities because each polymer absorbs a large amount of solvent, known as its 'hydrodynamic volume'. The hydrodynamic volume is frequently quantified by its viscosity (dL/g). The intrinsic viscosity of polysaccharides is directly related to many rheological properties. Solubility of polysaccharides in water is very low under normal conditions. The cause of this is the low degree of freedom for motion of stiff chains that disintegrates, which leads to minor increases in entropy (Morris et al., 1981). Enthalpy factors that favour the solid state, such as hydrogen bonding may not lead to dissolution at equilibrium if the free energy is not decreased.

$$\Delta G = \Delta H - T\Delta S$$

Here, ΔG is free energy, ΔH is enthalpy, T is temperature and ΔS denote entropy.

#### 4.3.1.3 Shape of Chain

The rigidity of monomer units results in polymers that have different shapes, as the geometries surrounding glycosidic bonds are a crucial factor. The angle of torsion establishes the orientations of adjacent residue groups, which together define the polymer conformation throughout the chain. These conformations are temporally stable in the solid state, but in solution, local conformation undergoes significant

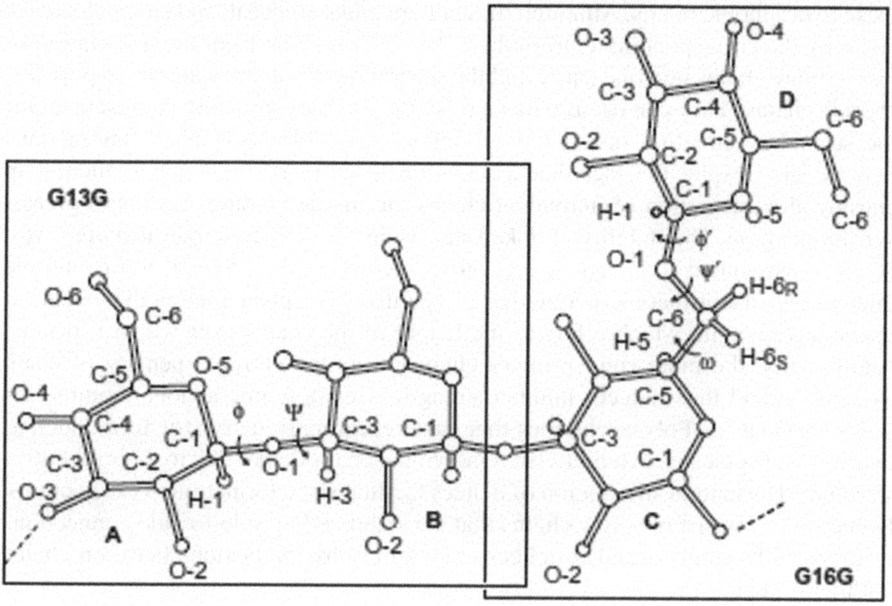


**Fig. 4.3** The shape of (1,3)-β-glucan chain (Palleschi et al., 2005)

changes, requiring investigation into average chain shapes. Certain types of connections between carbohydrate units create defined geometries that are determined by the location of the bonds involved. Misalignment of the 3 positions against the 1 position in the ring leads to misaligned chains with a (1,3)-β-glycosidic bonds, leading to 'twists' concerning the monomer unit. The underlying principle is that everything starts with an "I." The formation of helical chain shapes is a common feature among (1,3)-β-glucans that exhibit repetitive 'twist' bonds (Fig. 4.3).

#### 4.3.1.4 Intermolecular Interactions Promoting Chain Association

Due to the fact that many polysaccharide chains are soluble in aqueous solvents, the possibility of association with other chains must be considered. The process of association or aggregation is observed in equilibrium thermodynamics when the free energy of the associated state is lower than that of its dissolved state. However, it is also necessary to consider whether there are also structures that are kinetically trapped. If a state, such as an association form, is stabilized and does not contain the lowest level of free energy, it may be incapable of transitioning to correspondingly lower energy states due to high activation energy barrier, thus becoming kinetically trapped. A hydrogen bond exchange reaction is the main process that stabilizes interactions between uncharged polysaccharides in an enthalpic manner. 'Hydrogen bonds' are formed in this process, where these bonds between polysaccharides and solvent water are less energetically advantageous than the hydrogen bond between these hydrophobic bonds. Although these interactions are weak and cannot be overcome by the corresponding entropy loss. The later may be from the association of two or more chains into one entity and the separation of solvent water from polysaccharide chains. They can occur when a repeating primary structure is present and if the secondary structure becomes well-defined, multiple weak chain interactions may occur. Despite the fact that association leads to a significant reduction in entropy, the integration of individual chains into unified entities results in greater synchronization. The stability of linked states with weak interactions also improves. A stable associated state can be created by increasing the probability of multiple interactions in a primary structure that is repetitive. The phenomenon that occurs at cooperative length, which refers to the length of the chains over which it occurs. Additionally, the more rigid primary chain lowers the entropic penalty of chain association and thus reduces number/strength of entopic interactions required for stable association. Polysaccharides that have repeating primary structures, such as many (1,3)-β-glucans, exhibit twisted individual chain shapes due to their repetitive structure. The natural stimulation of helices facilitates the formation of connections. Helixes may consist of single chains that are stabilized by side-to-side connections or they can be multi stranded helices caused only by interactions between chains within the chain.

#### 4.3.1.5 Formation of Networks

For the production of strong three-dimensional networks, it is essential to establish stable intermolecular connections in aqueous surroundings, as demonstrated by gels produced under highly hydrated conditions. Conversely, if the molecular mobility of an entire polysaccharide chain is restricted and only rigid inter-chain associations are established, then there is an entropic cost. Therefore, obtaining the minimum free energy usually involves a balanced combination of enthalpically stabilized association zones (also known as 'junctions') created from cooperative interactions that span over definite lengths and regions of molecular flexibility that contribute to elastic stability and thereby contribute directly to overall determinants of total protein content or molar mass/weight. When there are multiple inter-chain associations per molecule and entropically stabilized interjunction regions between different molecules, a network structure is created with the help of this phenomenon that can be traced back to an intermediate linkage point.

## 4.4 Methods of Determination

### *4.4.1 X-Ray Diffraction*

The currently employed approach for detailed determination of the conformation of polysaccharide chains involves the use of X-ray or electron fiber diffraction. Due to the inherent lack of spontaneous crystallization in these polysaccharides, it becomes necessary to meticulously prepare fibers and apply treatments to enhance crystallinity in order to create a crystal lattice capable of diffracting X-rays. However, these treatments often deviate from real conditions, thereby necessitating the development of methodologies that ensure similar conformations under varying environments.

Upon obtaining a diffraction pattern that contains maximum information, unit cell indexing and molecular structure determination are carried out using standard protocols. However, the experimental data often lack the necessary information to explicitly determine the three-dimensional structure. Consequently, molecular modeling techniques are employed to predict energetically favorable conformations, thereby reducing the pool of potential structures. By continually refining fits to the experimental data through model building and refinement, three-dimensional structures can be proposed in order to minimize residual errors. In certain cases, the crystallization of oligosaccharide models has proven advantageous, as these lower molecular weight systems more readily form single crystals, thereby allowing for high-resolution analysis.

The advent of high-power X-ray sources has facilitated the analysis of crystals on a micron scale, eliminating the need to cultivate millimeter-sized crystals for conventional X-ray diffraction. As X-ray fiber diffraction is a specialized technique that is only available in some laboratories worldwide, powder diffraction "fingerprints" are often utilized to characterize the presence of crystalline conformational polymorphs in polysaccharide systems. This method, which employs low-power X-rays and standard laboratory equipment, offers a more accessible alternative to advanced techniques such as fiber diffraction and crystallography.

### 4.4.2 Solid State NMR

Solid-state Nuclear Magnetic Resonance (NMR) has emerged as a valuable tool alongside X-ray diffraction methods for the analysis of molecular conformations in polysaccharides. A particularly notable technique in this field is $^{13}C$ Cross-Polarization Magic Angle Spinning (CPMAS) NMR. While $^{13}C$ NMR is less sensitive than $^{1}H$ NMR, it offers a significant advantage by providing a wider range of chemical shifts for polysaccharides (approximately 50 ppm compared to 3 ppm for $^{1}H$). The cross-polarization technique enables the transfer of energy from protons to carbons, selectively exciting the rigid components of the system and enhancing the sensitivity of carbon spectra. Simultaneously, magic angle spinning (MAS) averages the chemical shifts for different molecular orientations within the sample, achieving an effect similar to that in solution-state NMR.

Solid-state NMR allows for the examination of individual atomic nuclei, providing insight into structure at a short-range level (typically sub-nanometre scale). In contrast, X-ray diffraction requires a minimum distance of approximately 10 nm and numerous unit cells in crystalline register for signals to be observed. Therefore, $^{13}C$ CPMAS NMR provides a means to probe molecular order at the sub-crystalline level and can detect signals from non-ordered or crystalline solid materials, offering an overview of solid-state structures. However, the assignment of signals often requires referencing primary conformation information obtained from X-ray diffraction or crystallography (Tan et al., 2007). The interpretation of $^{13}C$ CPMAS NMR involves recording spectra for samples with characteristic powder diffraction patterns of known crystalline polymorphs. These reference spectra aid in quantifying the degree of molecular order within a polysaccharide sample. Additionally, $^{13}C$ CPMAS NMR extends its applicability beyond dry solid samples to the investigation of wet solids, gels and other soft solid forms, detecting only the segments within the sample that exhibit "solid-like" behaviour. Complementary NMR techniques, such as $^{13}C$ Single Pulse Magic Angle Spinning (SPMAS), can detect more mobile molecular segments as solution-like signals, facilitating the description of conformation in terms of polymer flexibility, known as mobility-resolved spectroscopy.

### 4.4.3 Microscopy

Both electron microscopy and atomic force microscopy (AFM) have provided invaluable insights into the behaviour of (1,3)-β-glucans under highly diluted conditions, enabling the individual observation of molecular species. Thorough examinations, particularly focused on dilute solutions of schizophyllan, have uncovered a diverse array of persistent chains that exist independently or loosely aggregate when studied using electron microscopy on surface-deposited material. The observed rigidity of these molecular species is interpreted as an indication of the presence of triple helical conformations, supported by data from calorimetry and additional evidence. Significantly, electron microscopy has revealed the existence of cyclized structures, suggesting the supercoiling of triple helices—a phenomenon not previously reported for any other polysaccharide but observed in other inflexible multistranded helices such as DNA. The visualization of molecular species through electron microscopy requires precise preparation procedures, which may potentially influence the structures that are ultimately observed (Sletmoen et al., 2005). In contrast, AFM offers the advantage of directly monitoring structures, including those deposited on a molecularly smooth surface. By utilizing non-contact AFM, both linear and cyclized forms of schizophyllan have been identified, allowing for comprehensive investigations into the transformation between these two states as a function of thermal history.

### 4.4.4 Scattering and Sedimentation

A range of experimental methodologies, such as light scattering, small-angle X-ray scattering (SAXS), and sedimentation analysis produce valuable insights into the dimensions of molecules in different environments. These approaches consistently uncover triple helices as the primary conformational structure in solutions of (1,3)-β-glucans, as well as in tightly coiled macrocycles. Light scattering, typically carried out under highly diluted solution conditions, provides information on molecular dimensions, while SAXS offers additional details on longer-range structures, such as liquid crystals that form at higher concentrations.

An examination focused on cinerean, a (1,3)-β-glucan with side chains branching off the backbone at a ratio of approximately one in three units, investigated the conditions that facilitate the formation of liquid crystals. This phenomenon entails the presence of high concentrations of rigid, rod-like triple helices and a limited molecular weight achieved through the process of sonication. The study validated the existence of a triple helix with a diameter of 1.9 nm and demonstrated that the phase diagram agreed quantitatively with established theories of liquid crystallinity.

### *4.4.5 Fluorescence Spectroscopy*

The most important pathway for the study of (1,3)-β-glucans involves the utilization of vibrant color particles through meticulous binding. Aniline Blue, a readily accessible pigment, is utilized as a chemical probe for (1,3)-β-glucans, although its active component consists of a minor compound, sodium carbonylbis (4-(phenyleneamino) benzenesulphonate). While the luminescent photophysics of the active compound are well-documented, the precise mechanism by which it binds to (1,3)-β-glucans remains uncertain. Conditions with elevated pH, which are sufficient to disrupt the triple helices of (1,3)-β-glucans, lead to improved luminescence of color particles, suggesting a correlation with *Limulus amoebocyte* activation. Initially, it was believed that a single helix conformation was responsible for the binding, but subsequent spectroscopy data on luminescence resonance energy transfer (LRET) contradicted this hypothesis (Young et al., 2000). Rather, the LRET data supported the presence of an incompletely unwound triple helix with single chain binding sites. These findings contribute to the ongoing discussion regarding the conformational structures of (1,3)-β-glucans in predominantly aqueous environments, which are often associated with natural conditions. Previous studies indicated the existence of single helix configurations, but recent investigations on diluted solutions of (1,3)-β-glucans have generally linked triadic helix and random coil configurations. Despite uncertainties regarding the precise molecular interactions with the fluorescent component of Aniline Blue, this response remains crucial in histological examinations of (1,3)-β-glucans and serves as the basis for microtiter quantitative assays. Furthermore, it also holds implicit value as a method for assessing the distribution and quantity of (1,3)-β-glucans in food products (Ko & Lin, 2004).

### 4.4.6 X-Ray Diffraction of (1,3)-β-Glucans

There aren't enough X-ray diffraction studies of (1,3)-β-glucans that may be used as a baseline for interpreting conformational characteristics in a variety of environmental settings. Based on density and stereochemical considerations in conjunction with limited X-ray diffraction data, a triple helix structure is proposed. Deslandes et al. (1980) reported diffraction patterns with increased intensity and resolution after extruding a 10% DMSO solution of curdlan into methanol. After being formed, the fibers were evacuated to yield an anhydrous sample and annealed under tension in the presence of water at 145 °C. It was demonstrated that the final diffractogram shared characteristics with earlier, lower-quality diffraction patterns despite the fake nature of the fabrication method. Deslandes et al. (1980) proposed a precise molecular structure for a triple helical form of linear (1,3)-β-glucans based on X-ray analysis of the annealed anhydrous fiber and stereochemical model refining. This structure has remained the standard. It was suggested that the triple helices stabilize the crystalline structure by being made up of parallel strands with several hydrogen bonds between and within each triple helix. The crystalline unit cell was hexagonal with space group P63. Chuah et al. (1983) then combined powder diffraction patterns for paramylon granules with diffraction patterns from the hydrated annealed fiber to develop a model for the hydrated form of triple helical (1,3)-β-glucans. The resulting hexagonal unit cell structure was highly reliable and demonstrated that, despite the space group being different ($P_1$), the individual triple helices were virtually identical to the geometry suggested for the anhydrous form with the same stabilizing intra-helix hydrogen bonds. The high-resolution crystal structures of cellulose polymorphs that are currently available show that further advancement in pinpointing the precise locations of hydrogen atoms may be achievable by using neutron scattering with specific isotopic replacement of hydrogen with deuterium (Nishiyama et al., 2003). Nevertheless, there is evidence suggesting that the ordered conformation observed in (1,3)-β-glucans may not be limited to the triple helix. Gels that can be stretched into (weakly) crystalline structures that differ in diffraction patterns from both of the well-defined triple helical structures are generated when alkaline solutions of curdlan are neutralized. For this highly hydrated form, two structural interpretations have been put forth. According to Okuyama et al. (1991), one is based on a less tightly coiled triple helical structure, and the other is based on highly hydrated single helices. It was demonstrated that the highly hydrated version of curdlan's diffraction pattern, which has a low information content, is consistent with a variety of helical configurations, including several single, double, and triple helices.

### 4.4.7 Solid State 13 C NMR of (1,3)-β-Glucans

Owing to their various biological origins and existence as crystalline allomorphs, (1,3)-β -glucans were among the first polysaccharide families to be thoroughly investigated by solid state CPMAS $^{13}$C. NMR. The results showed diverse spectral features depending on biological origin. The spectra of curdlan, paramylon, and the linear (1,3)-β-glucans, namely the C-3 resonance, were found to be sensitive to moisture content by Fyfe et al. (1984). At greater moisture levels, shorter lines were observed, which suggested a more well-defined conformation. NMR lines for the more crystalline paramylon were found to be smaller than those for curdlan. The investigation of hydration effects was expanded to include a variety of (1,3)-β-glucans with varying molecular weights by Saito et al. (1989). With three main forms identified that differed based on biological origin and sample treatment, the C-3 site was determined to be the most variable in chemical shift. These were known as "annealed," "hydrate," and "anhydrous," and they had distinct chemical shifts related to curdlan. C-3 chemical shifts for "hydrate" and "amorphous" forms vary between curdlan and and laminaran. All other (1,3)-β-glucans examined, with the exception of the "annealed" form, which appears to have similar spectra. "Amorphous" forms, such as anhydrous curdlan, are difficult to interpret in detail without the results of molecular orbital calculations to predict the effect of conformational parameters on $^{13}$C chemical shifts. These forms have broad spectral features, consistent with the lack of a well-defined repeating conformation. There seems to be agreement that the 'annealed' form corresponds to the triple helical structure characterized by X-ray diffraction. Anhydrous curdlan has limited crystallinity and broad NMR signals and the 'hydrate' form that has different diffraction patterns and NMR spectra.

However, it is also likely that the 'hydrate' form resembles to a conformation based on triple helices with either partially open 'frayed' ends (Young et al., 2000) or cyclized forms of triple helices with mismatches between strands producing regions of incomplete helix formation. In both these cases, segmental mobility within the loosely held (1,3)-β-glucan chains can be imagined to be satisfactory for visibility in solution state NMR experiments. This explanation would also be consistent with the enzymic synthesis of (1,3)-β-glucans under similar conditions leading to either the 'annealed' or 'hydrate' forms (Pelosi et al., 2006).

## 4.5 Conclusion

Beta-glucan is a predominant non-starch polysaccharide composed of linear chains of β-d-glucose linked by β-(1 → 3), (1 → 4), and/or (1 → 6)-D-glycosidic linkage either in branched or unbranched form. The major source of beta-glucan involves cereals, microorganism, mushroom, lichens, and seaweeds. Beta-glucan shows structural variability based on its source. Based on the source and extraction method, beta-glucan can exist in a range of conformations such as random coils, helices (single, double, or triple), rod like shape, and wormlike shapes. The molecular weight of beta-glucan varies from 102 to 106 Da, depending upon the origin. Each type of beta-glucan encompasses different molecular backbone and level and/or length of branching which eventually affects its water solubility, ability to form aggregates, conformation (shape, size, and structure), molecular weight, techno-functional properties, and its physiological impact. The application of recent advanced techniques of analysis like FTIR, NMR, fluorescence spectroscopy, XRD and microscopy has made it possible to reveal and explore the various hidden characteristics of the beta-glucan to further its applications in food industry.

## References

Akima, K., Arika, T., & Amemiya, K. (1985). Purification, physicochemical characterization, and antitumor-activity of a cancer-associated human-serum protein that is increased by treatment with schizophyllan, an antitumor polysaccharide. *Japanese Journal of Cancer Research, 76*, 541–548.

Archibald, A. R., Cunningham, W. L., Manners, D. J., Stark, J. R., & Ryley, J. F. (1963). Metabolism of the protozoa, X. The molecular structure of the reserve polysaccharides from *Ochromonas malhamensis and Peranema trichophorum. Biochemical Journal, 88*, 444–451.

Blaschek, W., Kasbauer, J., Kraus, J., & Franz, G. (1992). *Pythium aphanidermatum*–culture, cell wall composition, and isolation and structure of antitumor storage and solubilized cell wall (1-3), (1-6)- β -D-glucans. *Carbohydrate Research, 231*, 293–307.

Bruneteau, M., Fabre, I., Perret, J., & Michel, G. (1988). Antitumour active β -D-glucans from *Phytophthora parastica. Carbohydrate Research, 175*, 137–143.

Carbonero, E. R., Montai, A. V., Woranovicz-Barreira, S. M., Gorin, P. A. J., & Iacomini, M. (2002). Polysaccharides of lichenized fungi of three *Cladina* spp.: significance as chemotypes. *Phytochemistry, 61*, 681–686.

Chang, Y. W., & Lu, T. J. (2004). Molecular characterization of polysaccharides in hot-water extracts of *Ganoderma lucidum* fruiting bodies. *Journal of Food and Drug Analysis, 12*, 59–67.

Chizhov, A. O., Dell, A., Morris, H. R., Reason, A. J., Haslam, S. M., McDowell, R. A., Chizhov, O. S., & Usov, A. I. (1998). Structural analysis of laminarans by MALDI and FAB mass spectrometry. *Carbohydrate Research, 310*, 203–210.

Chuah, C. T., Sarko, A., Deslandes, Y., & Marchessault, R. H. (1983). Packing analysis of carbohydrates and polysaccharides. Part 14. Triple-helical crystalline structure of curdlan and paramylon hydrates. *Macromolecules, 16*, 1375–1382.

Cordeiro, L. M. C., Stocker-Woergoetter, E., Gorin, P. A. J., & Iacomini, M. (2004). Elucidation of polysaccharide origin in *Ramalina peruviana* symbiosis. *FEMS Microbiology Letters, 238*, 79–84.

Cui, W., Wood, P. J., Blackwell, B. A., & Nikiforuk, J. (2000). Physicochemical properties and structural characterization by two-dimensional NMR spectroscopy of wheat β -D-glucan – Comparison with other cereal β -D-glucans. *Carbohydrate Polymers, 41*, 249–258.

Demirbas, A. (2005). β-Glucan and mineral nutrient contents of cereals grown in Turkey. *Food Chemistry, 90*(4), 773–777.

Deslandes, Y., Marchessault, R. H., & Sarko, A. (1980). Packing analysis of carbohydrates and polysaccharides triple-helical structure of (1-3)-beta-D-glucan. *Macromolecules, 13*, 1466–1471.

Fontaine, T., Simenel, C., Dubreucq, G., Adam, O., Delepierre, M., Lemoine, J., Vorgias, C. E., Diaquin, M., & Latge, J. P. (2000). Molecular organization of the alkali-insoluble fraction of *Aspergillus fumigatus* cell wall. *Journal of Biological Chemistry, 275*, 27594–27607.

Fry, S. C., Mohler, K. E., Nesslrod, B. H. W. A., & Franlova, I. (2008). Xyloglucan endotransglucosylase,a novel wall remodelling enzyme from *Equisetum* (horsetails) and charophyte algae. *Plant Journal, 55*, 240–252.

Fyfe, C. A., Stephenson, P. J., Taylor, M. G., Bluhm, T. L., Deslandes, Y., & Marchessault, R. H. (1984). Hydration effects in the C-13 CP/MAS NMR-spectra of solid (1-3)-beta-D-glucans. *Macromolecules, 17*, 501–502.

Gidley, M. J., & Nishinari, K. (2009). Physico-chemistry of (1, 3)-β-glucans. In *Chemistry, biochemistry, and biology of 1–3 beta glucans and related polysaccharides* (pp. 47–118). Academic.

Gummadi, S. N., & Kumar, K. (2005). Production of extracellular water insoluble β-,3-glucan (curdlan) from *Bacillus* sp SNC07. *Biotechnology and Bioprocess Engineering, 10*, 546–551.

Kenyon, W. J., Esch, S. W., & Buller, C. S. (2005). The curdlan-type exopolysaccharide produced by *Cellulomonas flavigena* KU forms part of an extracellular glycocalyx involved in cellulose degradation. *Antonie Van Leeuwenhoek, 87*, 143–148.

Kiss, J. Z., Vasconcelos, A. C., & Triemer, R. E. (1988). The intramembranous particle profile of the paramylon membrane during paramylon synthesis in *Euglena* (Euglenophyceae). *Journal of Phycology, 24*, 152–157.

Ko, Y. T., & Lin, Y. L. (2004). 1,3-beta-Glucan quantification by a fluorescence microassay and analysis of its distribution in foods. *Journal of Agricultural and Food Chemistry, 52*, 3313–3318.

Kollar, R., Reinhold, B., Petrakova, E., Yeh, H., Ashwell, G., Drgonova, J., Kapteyn, J., Klis, F., & Cabib, E. (1997). Architecture of the yeast cell wall – β (1 → 6)-glucan interconnects mannoprotein, β (1 → 3)-glucan, and chitin. *Journal of Biological Chemistry, 272*, 17762–17775.

Lee, J., Mullins, J. T., & Gander, J. E. (1996). Water-soluble reserve polysaccharides from Achlya are (1→ 3)-β-glucans. *Mycologia, 88(2), 264–270.*

Lazaridou, A., Biliaderis, C. G., Micha-Screttas, M., & Steele, B. R. (2004). A comparative study on structure–function relations of mixed-linkage (1→ 3), (1→ 4) linear β-d-glucans. *Food Hydrocolloids, 18*(5), 837–855.

Morris, E. R., Cutler, A. N., Ross-Murphy, S. B., Rees, D. A., & Price, J. (1981). Concentration and shear rate dependence of viscosity in random coil polysaccharide solutions. *Carbohydrate Polymers, 1*(1), 5–21.

Nakashima, A., Yamada, K., Iwata, O., Sugimoto, R., Atsuji, K., Ogawa, T., & Suzuki, K. (2018). β-Glucan in foods and its physiological functions. *Journal of Nutritional Science and Vitaminology, 64*(1), 8–17.

Nishantha, M. D., Diddugoda, J., Nie, X., & Weining, S. (2018). Beta-glucan: An overview of its properties, health benefits, genetic background and practical applications. *Scholars Journal of Agriculture and Veterinary Sciences, 5*, 130–140.

Nishiyama, Y., Sugiyama, J., Chanzy, H., & Langan, P. (2003). Crystal structure and hydrogen bonding system in cellulose I α, from synchrotron X-ray and neutron fiber diffraction. *Journal of America Chemical Society, 125*, 14300–14306.

Okuyama, K., Otsubo, A., Fukuzawa, Y., Ozawa, M., Harada, T., & Kasai, N. (1991). Single-helical structure of native curdlan and its aggregation state. *Journal of Carbohydrate Chemistry, 10*, 645–656.

Palleschi, A., Bocchinfuso, G., Coviello, T., & Alhaique, F. (2005). Molecular dynamics investigations of the polysaccharide scleroglucan: First study on the triple helix structure. *Carbohydrate Research, 340*(13), 2154–2162.

Papageorgiou, M., Lakhdara, N., Lazaridou, A., Biliaderis, C. G., & Izydorczyk, M. S. (2005). Water extractable (1-3), (1-4)-β-D-glucans from barley and oats: An inter- varietal study on their structural features and rheological behaviour. *Journal of Cereal Science, 42*, 213–224.

Pelosi, L., Bulone, V., & Heux, L. (2006). Polymorphism of curdlan and (1-3)-beta-D-glucans synthesized in vitro: A C-13 CP-MAS and X-ray diffraction analysis. *Carbohydrate Polymers, 66*, 199–207.

Portilho, M., Matioli, G., Zanin, G. M., Faria de Moraes, F., & Scamparini, A. R. P. (2006). Production of insoluble exopolysaccharide of *Agrobacterium* sp. (ATCC 31749 and IFO 13140). *Applied Biochemistry and Biotechnology, 131*, 864–869.

Rout, D., Mondal, S., Chakraborty, I., Pramanik, M., & Islam, S. S. (2005). Chemical analysis of anew (1 → 3)-, (1 → 6)-branched glucan from an edible mushroom, *Pleurotus florida. Carbohydrate Research, 340*, 2533–2539.

Ruiz-Herrera, J., Elorza, M. V., Valentin, E., & Sentandreu, R. (2006). Molecular organization of the cell wall of *Candida albicans* and its relation to pathogenicity. *FEMS Yeast Research, 6*, 14–29.

Saito, H., Yokoi, M., & Yoshioka, Y. (1989). Effect of hydration on conformational change or stabilisation of (1-3)-beta-D-glucans of various chain lengths in the solid state as studied by high-resolution solid-state C-13 NMR-spectroscopy. *Macromolecules, 22*, 3892–3898.

Shivakumar, S., & Vijayendra, S. V. N. (2006). Production of exopolysaccharides by *Agrobacterium* sp CFR-24 using coconut water – A by-product of food industry. *Letters in Applied Microbiology, 42*, 477–482.

Sletmoen, M., Christensen, B. E., & Stokke, B. T. (2005). Probing macromolecular architectures of nano-sized cyclic structures of (1→ 3)-β-d-glucans by AFM and SEC-MALLS. *Carbohydrate Research, 340*(5), 971–979.

Smiderle, F. R., Carbonero, E. R., Mellinger, C. G., Sassaki, G. L., Gorin, P., & Iacomini, M. (2006). Structural characterization of a polysaccharide and a β -glucan isolated from the edible mushroom *Flammulina velutipes. Phytochemistry, 67*, 2189–2196.

Tan, I., Flanagan, B. M., Halley, P. J., Whittaker, A. K., & Gidley, M. J. (2007). A method for estimating the nature and relative proportions of amorphous, single, and double-helical components in starch granules by 13C CP/MAS NMR. *Biomacromolecules, 8*(3), 885–891.

Wang, M. C., & Bartnicki-Garcia, S. (1980). Distribution of mycolaminarans and cell wall β-glucans in the life cycle of Phytophthora. *Experimental Mycology, 4*, 269–280.

Wang, Y., Zhang, M., Ruan, D., Shashkov, A. S., Kilcoyne, M., Savage, A. V., & Zhang, L. (2004). Chemical components and molecular mass of six polysaccharides isolated from the sclerotium of *Poria cocos. Carbohydrate Research, 339*, 327–334.

Wang, L., Ye, F., Li, S., Wei, F., Chen, J., & Zhao, G. (2017). Wheat flour enriched with oat β-glucan: A study of hydration, rheological and fermentation properties of dough. *Journal of Cereal Science, 75*, 143–150.

Xu, C., Lv, J., Lo, Y. M., Cui, S. W., Hu, X., & Fan, M. (2013). Effects of oat β-glucan on endurance exercise and its anti-fatigue properties in trained rats. *Carbohydrate Polymers, 92*(2), 1159–1165.

Young, S. H., Dong, W. J., & Jacobs, R. R. (2000). Observation of a partially opened triple-helix conformation in 1→ 3-β-glucan by fluorescence resonance energy transfer spectroscopy. *Journal of Biological Chemistry, 275*(16), 11874–11879.

Zhang, X. F., Zhang, L. N., & Xu, X. J. (2004). Morphologies and conformation transition of lentinan in aqueous NaOH solution. *Biopolymers, 75*, 187–195.

Zhang, J., Yan, L., Liu, M., Guo, G., & Wu, B. (2021). Analysis of β-d-glucan biosynthetic genes in oat reveals glucan synthesis regulation by light. *Annals of Botany, 127*(3), 371–380.

# Chapter 5
# Physicochemical Properties of Beta-Glucans

**K. U. Anjali, A. Rajalechumi Kamatchi, and Sundaramoorthy Haripriya**

## 5.1 Introduction

Dietary fibers help maintain glucose homeostasis in pre-diabetic individuals or those with type II diabetes by lowering blood cholesterol, improving gut health, avoiding risk, and treating chronic ailments. Therefore, adding dietary fiber (DF) could make foods healthier, and lower in calories, cholesterol, and fat (Mejía et al., 2020). The worldwide dietary recommendations frequently advocate for large intakes of fiber because of the health benefits accredited to dietary fiber. However, 25–38 g of fiber per day, advised for adults fails to indicate particular guidelines on the kind of fiber to devour. Indeed, human diets contain a wide diversity of dietary fiber origins, such as cereals, fresh produce, nuts, seeds, and legumes which have significantly different physicochemical features (Henrion et al., 2019).

Beta-glucans are dietary fibers, a long-chain glycan, comprising D-glucose monomers, interlinked with glucosidic bonds (Maheshwari et al., 2017). A wide range of biofunctions, comprising anti-tumor, immune-regulating, anti-aging, and anti-inflammatory effects, are exhibited by β-glucan. Beta-glucans have garnered interest for several years due to their unique physical and molecular characteristics. Beta-glucans from various origins and with various molecular weights (MW) have differences in their biological effects. The baker's yeast beta-glucan has β-(1 → 3) and (1 → 6) connections, and cereal-sourced beta-glucans are glycans of carbohydrate remnants with β-(1 → 3) and β –(1 → 4) connections. Additionally, curdlan is a (1 → 3) beta-glucan, cellulose is a (1 → 4) beta-glucan, and lichen beta glucan holds either (1 → 3) (1 → 4) or (1 → 6) linkages. Figure 5.1. depicts the different sources and the branching of glucan chains. Fungus beta-glucan has demonstrated

K. U. Anjali · A. R. Kamatchi · S. Haripriya (✉)
Department of Food Science and Technology, Pondicherry University, Puducherry, India



M. Bashir Mir, S. A. Mir (eds.), *Beta-Glucan: Sources, Properties and Applications*,
https://doi.org/10.1007/978-3-031-95788-8_5

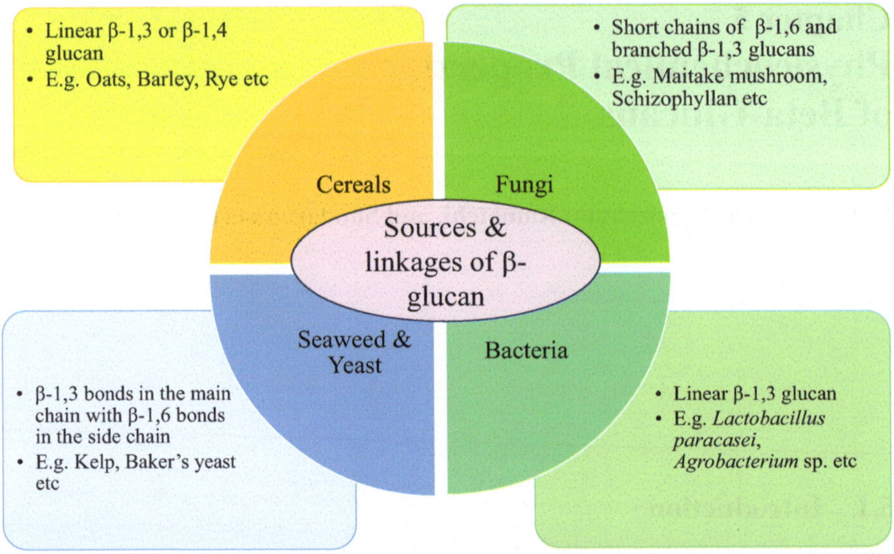


**Fig. 5.1** Sources and branching of β-glucan (Zhu et al., 2016; Nakashima et al., 2018 & Du et al., 2019)

efficiency as an immunity stimulant and anti-cancer agent. Cereal beta-glucan helps to reduce blood sugar and cholesterol levels in the human body (Zhu et al., 2016).

The presence of multiple hydroxyl clusters in the β-glucan fosters the linking with reactive groups of other molecules and alters the solubility, conformation, and aggregation ability. The MW of the beta-glucan is influenced by the position, degree, and length of branching as well as the structure, shape, and size of the branches. These inherent features play a vital part in making the beta-glucan complete diverse functional roles in food products and also contribute to the biological functions in the living system (Kaur et al., 2019). The physical and physicochemical characteristics of beta-glucans hold considerable importance in the contexts of entrepreneurship and nutrition. The recognition of β-glucans as biopotent compounds during the past two decennium has escalated engrossment in these compounds. On the basis of their flow characteristics, beta-glucans have also been considered for use as food hydrocolloids. Under specific circumstances, β-glucans have demonstrated the gel-forming potential and increase the viscosity in liquids. Beta-glucans have the capability to improve the viscosity of various food formulations by acting as thickeners, or alternatively, they can be employed as fat substitutes to produce low-calorie foods. In addition, because of their advantageous physiological effects, β-glucans from cereals have recently been the subject of numerous attempts to increase their content in food formulations. So far, products like breakfast cereals, spaghetti, noodle dishes, baked items, dairy, and meat products have been successfully incorporated with ample fractions of beta-glucan from cereal. The linking arrangements, conformity, and molecular weight or distributions of beta-glucans generally govern their physical properties, including their solubility and rheological conduct in suspension and gel conditions (Lazaridou et al., 2007). However, the connections between structural characteristics and physiological reactions have only been relatively studied.

This chapter emphasizes the distinct physicochemical attributes of β-glucan and their relationship with various factors, and functionality.

## 5.2 Physicochemical Properties

### *5.2.1 Molar Ratio*

The molar ratio pertains to the proportion of the cellotriosyl and cellotetrosyl (DP3/DP4) units ensuing to the depolymerization of the beta-glucan. Cereals are distinguished by this ratio, which serves as a unique form of identification for their beta-glucan structure. The molar ratio varies according to the sources; for instance, it ranges from 3 to 4.5 for wheat, 1.8 to 3.5 for barley, 1.9 to 3.0 for rye, and 1.5 to 2.3 for oats (Mejía et al., 2020). The presence of the cellotetrosyl components and the elongated β-1,4 linkages is a fundamental factor responsible for the formation of gel and the elevated viscosity in cereal beta-glucans. The regions containing β-1,4 linkages are pliable with heightened intermolecular bonding which strengthens the connections between glucan chains. This imparts elevated viscosity by inducing greater resistance to the flow. In contrast, the sections with β-1,3 bonds are ascribed to the entanglements in the glucan chains (Ryu et al., 2012). The molar ratio greatly influences the functional characteristics of the beta-glucans. The beta-glucan formulations which have a high molar ratio resulted in less solubility (Table 5.1). Higher concentrations of cellotriosyl units lead to the densely packed configurations of the beta-glucan, consequently resulting in reduced solubility by limiting space for water molecules to interact with the beta-glucan (Schmidt, 2022). It was found that beta-glucans with elevated molar ratios exhibit increased resistance to β-glucanase (De Arcangelis et al., 2019). These findings of enzymatic degradation provide evidence to estimate and regulate the future application of beta-glucan in food formulations. Also, it is important to consider the concomitant reduction of solubility in the targeted applications. The molar ratio relies upon the agronomic conditions and the cultivar species of the cereals. Care should be taken while selecting the cereals for selective beta glucan food applications (Herrera et al., 2016).

### *5.2.2 Molar Mass*

The molar mass of the beta-glucan plays a crucial role in ascertaining the characteristics that influence both the physiological effect upon consumption and dietary functionality (Table 5.1). Specifically, molar mass is closely connected with the

**Table 5.1** Physicochemical properties of beta-glucans

| Sources | Property | Inferences | References |
|---|---|---|---|
| Cereals | Molar ratio | Higher molar mass reduces the solubility | (Schmidt, 2022) |
| Barley | Molar ratio | An increase in molar ratio enhances the β-glucanase resistance | (De Arcangelis et al., 2019) |
| Cereals | Molar mass | The molar mass of beta-glucan greater than 210 kDa exhibits better functionalities and health outcomes | (Whitehead et al., 2014) |
| Oats | WBC | Elevated temperatures reduce the WBC | (Oliveira et al., 2012) |
| Millets | WBC | Germination enhances the WBC | (Sharma et al., 2018) |
| Oat bran | FBC | Modification enhances the FBC | (De Souza et al., 2015) |
| Yeast | SI | An increase in the radiation dose resulted in the reduction of SI | (Hongjie Yuan et al., 2020) |
| Foxtail and kodomillet | SI | Germination elevated the SI | (Sharma et al., 2018) |
| Highland barley | Solubility | Solubility increased with elevated temperature and reduced molecular weight | (Li et al., 2021) |
| Oats | Solubility | Degradation elevated the solubility | (Kaur et al., 2019) |
| Cereals | Emulsifying capacity | Molecular weight positively affected the emulsifying capacity | (Karp et al., 2019) |
| Mushroom | Emulsifying capacity | An increase in the irradiation dose positively correlated with the emulsifying capacity | (Ashraf et al., 2015) |
| Oats | FC | Concentration influenced the FC | (Sukriti Singh & Maninder Kaur, 2018) |
| Highland barley | FC | Increased molecular weight and viscosity enhance the FC | (Li et al., 2021) |
| Oats | Gelling | A drop in the molecular weight elevated the gelation rate | (Skendi et al., 2003) |
| Wheat | Gelling | Molecular weight reduction enhanced the gelation rate | (Li et al., 2011) |

*WBC* water binding capacity, *FBC* fat binding capacity, *SI* swelling index, *FC* foaming capacity

viscosity of beta-glucans in food commodities, shaping their texture and overall quality. Moreover, while consumption of molar mass impacts the physiological traits of the colon by impacting absorption and digestion (Mejía et al., 2020). Comprehending the molar mass of β-glucan is vital in the food science and human physiology domains to understand its complex implications. The molar mass of the beta-glucans from cereals exhibits a considerable variation among the source itself. The molar mass of beta-glucan from barley ranges from 50 to 200 kDa as indicated by the European Food Safety Authority (Panel & Nda, 2011). The diversity in the molar mass is primarily caused due to the environmental variables. Significant molar mass fluctuations can be attributable to other factors such as extraction methods, purification, depolymerization, and analytical protocols employed for the molar mass calculations counting the detectors and standards (Mejía et al., 2020). The molar mass of the β-glucans sourced from oats and barley on an industrial scale falls under 200–300 kDa (Mikkelsen et al., 2013). The other factors emphasizing the complex features of the molar mass encompass the physical form of the beta-glucan within the plant tissue, the functional activity of the β-glucanase enzyme, and processing and preservation methods (Regand et al., 2011).

The molar mass of cereal beta-glucans has a pivotal role in human wellness. Especially the molar mass exceeding 210 kDa demonstrates better qualities and contributes to favorable health outcomes. Beta-glucans incorporated in the dry food matrices, especially in breakfast cereals exhibited remarkable lipid profile enhancements, cholesterol level reduction, and effective obesity control. This suggests that the molar mass of beta-glucan may play an essential role in achieving health care in the realm of lipid metabolism, cholesterol, and weight management (Whitehead et al., 2014). The molar mass also takes part in assessing the prebiotic potential of beta-glucan governing the microbiota growth in the colon. Invitro studies inferred that the beta-glucan hydrolysates below 172 kDa enhanced the population of *Bacteroides- Prevotella* within 24 h (Hamaker & Tuncil, 2014). The diversity in the molecular characterization correlated with their physicochemical, rheological, and health aspects needs to be thoroughly comprehended before the application of beta-glucan in food product development.

### *5.2.3 Water Binding Capacity (WBC)*

The beneficial and potential uses of polysaccharides, particularly in the food sector, are greatly influenced by their hydration properties. The water absorption, binding capacity, and swelling power are used to assess hydration properties. Several factors, including temperature, soaking period, the extraction process, and centrifugation, determine the water-holding capacity of a given amount of dry polysaccharide. In food applications where water retention is desirable, such as in sausages and hamburgers, it offers insights into the capacity of polysaccharides to bind and retain water. The WBC is vital for preserving the fleshiness of the finished food product (Borchani et al., 2016).

The water binding capacity or water retention capacity exhibited by beta-glucans is a result of the mechanical stress offered during centrifugation. This beneficial feature can affect the storage life of the food product and can be utilized in diverse food industry applications. The WBC of the beta-glucan can differ on the sources and extraction procedures. The beta-glucan from barley extracted using hot water demonstrated higher WBC compared to the acid, alkali, and enzymatic extraction processes (Ahmad et al., 2009).

The temperature at which the beta-glucan is treated also has a notable influence on the WBC. The WBC of beta-glucan decreases at higher temperatures, this may be attributed to the abrasion of their continued conformations. The volume of water that oat β-glucan retains is significantly influenced by the temperature at which the fiber is processed. Higher temperatures have been found to decrease the water binding capacity, which may be ascribed to the physical impairment of the long chain of the fiber. The beta-glucan can absorb water quickly and stabilize rapidly since it has significant hydrophilic traits (Oliveira et al., 2012). This feature increases the water retention capacity of food products, mainly when frozen bakery goods are produced, because of its ability to resist water evaporation in the thawing process.

Additionally, it has been discovered that following germination, the WBC of β-glucan extracted from millets dramatically increases (Sharma et al., 2018). The numerous sites for water retention in seed composition are accountable for this improved WBC. In general, the WBC of beta-glucan is a crucial feature in food applications, and it might differ based on the source, extraction procedures, and treatment circumstances. Understanding and improving the WBC of the β-glucan enhances its use in food production to minimize water loss during thawing, improving the storage life of the food products and benefiting the utilization in a broad range of food applications.

### *5.2.4 Fat/Oil Binding Capacity (FBC/OBC)*

A beneficial property of β-glucan that makes it an appealing ingredient for functional foods for weight management and obesity treatment is its fat/oil binding capacity (FBC) associated with dietary fat digestion and absorption (Abdel-Haleem et al., 2020). In addition, beta-glucan's oil binding capacity (OBC) is essential for giving food commodities a palatable texture, flavor, and mouthfeel.

The molecular weight, types, and extent of modifications of beta-glucan impact its fat-binding capacity. Chemical modifications like acetylation often lead to lower FBC relative to native β-glucan. Enzymatic modifications can improve the FBC of the beta-glucan when collated to the original or native form. Additionally, it has been proved that beta-glucan with a smaller molecular weight showcases higher FBC (De Souza et al., 2015). The source of beta-glucan also has a notable impact on its FBC. Numerous studies have revealed varying FBC values with various sources of β-glucan. The FBC for oat beta-glucan is in the span of 3.9 and 11.4 g oil/g (Abdel-Haleem et al., 2020). The FBC values for oat β-glucan hydrolysates ranged from 6.72 to 11.36 g/g (Bae et al., 2009), for mushroom beta glucan it ranged from 5.38 to 6.65 g/g (Ashraf et al., 2017) and that of oat beta-glucan and its oligosaccharides ranged from 1.97 to 5.03 g/g (Qin et al., 2021). Contrary to expectations, the FBC of beta-glucan substantially rose, adhering to degradation, and it turned out that the FBC of hydrolyzed beta-glucan was more substantial than that of oxidatively degraded beta-glucan. This shows that the degradation process enhances the fat-binding ability and that the hydroxyl groups within the β-glucan are considered a vital factor for FBC over the carbonyl and carboxyl groups.

The distinctions in FBC emphasize the significance of considering aspects including beta-glucans origin, modifications, and molecular weight when utilizing it as a fat-binding component. The features mentioned earlier make β-glucan a promising ingredient in creating functional foods to treat obesity and aid in weight management. It is relevant to mention that even though the FBC of β-glucan exhibits affirmation, more investigation is required to fully comprehend the underlying mechanisms and assess its efficiency in food compositions.

### *5.2.5 Swelling Index (SI)*

Swelling Index/power is the measure of the swelling volume, which can be connected to the cholesterol-diminishing impact of the dietary fiber, since these swollen dietary fibers may restrict the digestion of nutrients and retention of digestive hydrolysates (Bae et al., 2009).

The swelling index of beta-glucan is a province of MW and is influenced by modification methods. The chemically modified acetylated β-glucan exhibited increased swelling power by 84% in contrast to the native beta-glucan. In contrast, the physically modified irradiated beta-glucan had a declined swelling power with an increased irradiation dose (Hongjie Yuan et al., 2020). Germination-influenced swelling power of beta-glucan obtained from the millets (Table 5.1). The swelling index of foxtail millet improved from 1.32 g/g to 1.67 g/g and from 2.54 to 2.99 g/g for Kodo millet after germination (Sharma et al., 2018). Extrusion and drying substantially improved the swelling index of the beta-glucan due to the emergence of capillaries and porous structures(Zhang et al., 2011).

The SI of oat beta-glucan is regulated by various degradation methods, including acid degradation and oxidation treatment (Sun et al., 2020). The swelling ability of oat beta-glucan is dramatically reduced after degeneration. The structural alterations that take place in oat beta-glucan may be the cause of the decrease in SI following degradation. The β-glucan conformation may disintegrate as a result of acid degradation and oxidation treatment, which will lessen the capacity of the β-glucan to expand when it comes in contact with water. This process results from the partial breakdown of oat beta-glucan, which incites a fall in MW. The capacity of the beta-glucan chain to retain moisture is reduced due to breakage in the glycosidic chain that happens during the process. These findings imply that food preparation and storage may impact oat beta-glucans SI, particularly when acid degradation is involved. These factors are to be considered while using oat β-glucan in culinary applications because the degradation processes affect the functional attributes of the substance.

### 5.2.6 Solubility

The polysaccharides exhibit a notable attraction for the water particles owing to the existence of numerous hydroxyl groups. The solubility of beta-glucan is a vital factor in influencing functional properties like stability, emulsifying characteristics, drug delivery, and membrane-forming features (Du et al., 2019). The propensity of β-glucan to get dissolved in water at a specific temperature and pressure to produce a homogenous hydrocolloid mixture is referred to as its solubility. Even though β-glucans are categorized as soluble fibers, their molecular variability can influence their solubility. Numerous variables, such as time, temperature, pH, and the specific source of the beta-glucan, could influence solubility. In general, β-glucan derived from oats is more water soluble than beta-glucan derived from barley. The beta-glucan from various sources has varying molecular structures and compositions, which leads to the difference in their solubility.

The solubility tends to decrease when the molar ratio rises or when there is a higher concentration of beta-glucan in water. The solubility of beta-glucan can also be impacted by the other biobased polymers in the exterior layer of the cell and their arrangement and interactivity. Due to the complex cell wall matrix of beta-glucan, it may take a longer time for the water to interact and solubilize the beta-glucan molecules by altering the mechanical resistance and permeability of the cell wall (Mejía et al., 2020). The solubility, chain conformation, and rheological characteristics of beta-glucans in solution are influenced by their size and structural characteristics. Cereal β-glucans have chemical properties that enable them to adopt extended, flexible chain conformations and make them soluble in water. They also have segments that resemble cellulose, increasing the molecule's rigidness in the solution. Moreover, due to powerful hydrogen bonding along the segments, β -glucans with nearby β -(1 → 4), connections tend to congregate between these chains, and this aggregation causes a decrease in solubility. The solubility and flexibility of the beta-glucans are increased by the presence of β -(1 → 3) connections, which interfere with the constancy of the β -(1 → 4) linkage series (Lazaridou et al., 2007).

Temperature, source, modification, and molecular weight affect the solubility of the beta-glucan. The highland barley bran β-glucans treated at various temperatures demonstrated a significant increase in solubility when their molecular weight declined (Table 5.1). Furthermore, the solubility index increases logarithmically concerning temperature. These results are consistent with the notion that higher solubility values are produced by β-glucans with lower molecular weights and higher solution temperatures (Li et al., 2021). Physical modifications like irradiation and high-pressure microjet homogenization improved the solubility of beta-glucans. Similarly, chemical modification of beta-glucans from various sources also displayed better solubility (Kaur et al., 2019).

### *5.2.7 Emulsifying Capacity (EC) and Emulsion Stability (ES)*

The hydrocolloid β -glucan can help to keep the emulsion systems stable by improving the consistency of the aqueous stage and lowering the agglomeration of air froth and oil drops. β-glucan from barley serves as a surfactant and is free of hydrophobic components. Temperature, pH, and concentration influence the emulsification properties of β-glucan in liquids. Higher β-glucan concentrations improve the emulsifying capacity by increasing the viscosity and reducing oil droplets. The best emulsifying capabilities of the β-glucan macromolecules are exhibited at a temperature of 45 °C. The stability of an emulsion is lowered by excessive heating, which reduces the viscosity of the solution. Emulsifying abilities perform best at neutral pH (7), then at acidic pH (4), and alkaline pH (9) conditions (Li et al., 2021). The impact of barley beta-glucan on emulsification showed that emulsions made with 4% barley β-glucan displayed the maximum stability throughout a two-week storage duration compared with emulsions made with oat beta-glucan, lecithin, and pectin. The emulsions made with barley β-glucan were more stable because of the high MW of the glycan and their capacity to form gel-like polymer networks (Karp et al., 2019). The beta-glucan with higher molecular weight stabilizes the oil in water emulsions primarily by elevating the viscosity of the continuous phase. In contrast, the low MW beta glucans influence the ES through the emergence of networks in the continuous phase (Lazaridou & Biliaderis, 2007). This highlights the significance of MW in regulating the stability of emulsions

Physical modifications like irradiation positively influence the emulsifying capacity and stability of the beta-glucan (Table 5.1). The impact of γ-irradiation on the beta-glucan sourced from button mushrooms found that increasing irradiation dosage improved the emulsifying capacity and emulsion stability. The maximum EC of 74.02% was found at a radiation dose of 50 kGy (Ashraf et al., 2015). Three varieties of mushroom β-glucans, *Agaricus, Pleurotus, and Coprinus*, were examined for their emulsion stability and capacity. The emulsion capacities of the beta-glucan from *Agaricus, Pleurotus, and Coprinus* were found to be 64.260.01%, 65.350.03%, and 65.470.08%, and the corresponding emulsion stability values were discovered to be 94.640.96%, 96.730.02%, and 97.680.20%, respectively. Higher percentages denote the greater emulsion capacity, and these values represent the capability of the beta-glucans to balance oil-in-water emulsion (Ashraf et al., 2017). Emulsion stability is the potential of the emulsion to withstand phase separation or breakdown over time. Higher stability percentages imply that the emulsion will be stable for a longer period.

### *5.2.8 Foaming Capacity (FC) and Foam Stability (FS)*

Foam is a multiphase network comprised of a continuous phase, which is a liquid, and a dispersed phase, which is typically an air or gas bubble. The dispersed phase is enclosed and surrounded by the continuous phase. Since polysaccharides are hydrophilic molecules, they have a strong affinity towards water. Polysaccharides typically stay in the aqueous subphase of a foam system due to their hydrophilic nature. Instead of forming a different phase within the foam structure, they dissolve or disperse in the liquid phase. Due to their characteristics, polysaccharides often serve as stabilizers and thickeners in colloidal systems, especially foams (Ashraf et al., 2015). Foams may arise when β-glucan is whipped or otherwise stirred. The structural changes that β-glucan exhibits after whipping or mechanical agitation enable it to create a stable network within the foam that aids in trapping and holding smaller solute particles (Li et al., 2021). This trait is advantageous in food manufacturing as it boosts the foaming capacity and foam stability, which may enhance food commodities' texture, appearance, and mouthfeel.

Foaming characteristics of the β-glucan isolated from the same source may vary with the different varieties. The oat β-glucan extracted from distinct Indian oat cultivars exhibited a variation in the foaming capacity from 20% to 48%, with the highest score obtained by the cultivar OS-7 and the lowest foaming capacity found in the OL-10 variety. The higher foaming capacity of the OS-7 is ascribed to the increased concentration of surface-active ingredients in this variety that contribute to the development and stabilization of the foam. The foam stability of the β-glucan dispersion degraded over time. An initial range of foam stability from 87.87–96.87% was reduced to 78.5–90.0% after 120 min of storage, indicating the foam dispersed upon storage (Sukriti Singh & Maninder Kaur, 2018).

Modification of β-glucan also has a substantial influence on their foaming characteristics. The FC and FS of irradiated yeast β-D-glucan peaked at 2.93% and 5.28%, respectively, with radiation exposure of 50 kGy. The two possible mechanisms for improved foam stability are as follows; (a) the foam stability involves the electrical multilayers produced by dipolar surfactants and the Vander Waals forces allying the molecules in the foam. These forces support the cohesive qualities of the foam, limiting bubble coalescence and preserving the structural integrity; (b)The network of polysaccharide molecules that develops within the foam is another possible explanation for this improved foaming characteristics. This network aids in sustaining the interfacial film between air and water by reinforcing the air bubbles with a strong film. The network created by polysaccharides cements the foam structure and slows the coalescence of bubbles, thus improving the foam's stability (Khan et al., 2016).

Foaming properties can be controlled by various parameters such as pH, temperature, concentration, and extraction methods (Table 5.1). Moderate heating and neutral pH environments often favored higher foaming capacity and stability, but extreme pH levels and excessively high temperatures had a contrary effect. Higher viscosity and molecular weight raised the foam stability but showed a negative correlation as they impeded foam generation associated with the foaming capacity and stability (Li et al., 2021). The beta-glucan obtained from raw and germinated millets found that germination increases solubilized proteins and fiber content, improving the FC and FS of millet beta-glucan extracts. The presence of solubilized proteins reduces surface tension and facilitates foam formation, while the protein film formed around air bubbles enhances foam stability (Sharma et al., 2018).

### *5.2.9 Gelling Characteristics*

The utilization of β-glucan as a food hydrocolloid relies predominantly on the rheological properties, especially the gelling capability and viscosity. Thus they can be incorporated to enhance the consistency and appearance in food preparations by employing it as a thickening agent. Moreover, they serve as fat mimics in the development of low-calorie foods with intensified sensory and nutritional attributes (Lazaridou & Biliaderis, 2007). The gelling properties of the β-glucan rely upon the MW (Table 5.1). The beta-glucan isolated from different oat varieties exhibited gelling potential. The beta-glucan with a lower MW displayed a higher gelation rate than the higher MW (Skendi et al., 2003). β-D-glucan with lower MW has a greater propensity to form gels than its counterpart. The enhanced mobility and distinctive structural patterns of the low molecular β-D-glucan facilitate the facile emergence of junction zones and a robust 3D matrix which accelerates the gelation rate (Du et al., 2019). The β-D-glucan from wheat presented augmented gelation rates as the molecular weight decreased owing to the elevated mobility and the reduced spatial obstacles (Li et al., 2011). A decline in the gelation time was observed in the water-soluble β-glucan from barley characterized by lower molecular weight emphasizing the impact of lower MW on the gelation rate contributing to the rapid gelling process (Vaikousi et al., 2004)

Temperature and modification methods significantly affect the gelling characteristics of the cereal beta-glucan. Investigations on the impact of temperature on the gelling ability of the cereals revealed that the optimum gelling temperature of the beta-glucans is around 37 and 57 °C respectively for oats and barley. The beta-glucans modified chemically by oxidation can produce gel at an equal temperature but result in softer gels (Lante et al., 2023). The weaker gel formation can be attributable to the polymer degradation induced by the ascorbic acid or hydrogen peroxide treatment. The modification of beta-glucan employing hydrogen peroxide is found to be an effective approach for regulating and optimizing the gelling potential. The method is considered to be economical and safe demonstrating a viable means to modify beta-glucan for flexible utilization in food manufacturing (Zhang et al., 2023). The partial hydrolysis of the oats β-D-glucan with cellulase decreased the MW and yielded the production of additional elastic gels with a well-built junction zone compared to the partial hydrolysates generated using lichenase (Table 5.1). This indicating the selection of appropriate hydrolytic enzymes also impacts the textural and rheology of the beta glucan-incorporated food products (Tosh et al., 2004).

### 5.2.10 Rheological Characteristics

Understanding the physical behavior of food products requires knowledge of rheology, which is the study of the flow and distortion of materials. Manufacturers can gain significant information about the texture, consistency, and stability of the finished food by inspecting the rheological attributes of food formulations. This knowledge guarantees product quality, improves processing conditions, and satisfies customer demands (Karp et al., 2019). The structure, molecular weight, polymer concentration, shear rate, and temperature affect the viscosity of the polysaccharide solution (Wang & Ellis, 2014).

The β-glucan isolated from different bran layers of high land barley exhibited a shear thinning behavior which has industrial benefits in reducing the energy consumption for pumping and mixing operations. The varying viscosities recorded in the β-glucan dispersions can be ascribed to the beta-glucan sources, molecular weight, and molecular structures (Table 5.2). Higher molecular weight beta-glucan frequently exhibits higher viscosities because of the well-built molecular chain entrapments and increased flow variance from lengthy polymers. The beta-glucan dispersion's flow behavior index (n) values dropped with improved molecular weight, indicating reduced shear-thinning behavior. The dynamic frequency sweep result of the β-glucan can be explained by the elastic component (G′) and the viscous component of the material (G″). Both the G′ and G″ of the β-glucan dispersion will be improved with the frequency, where the G″ scores of the beta-glucan dispersion were constantly lesser than the coinciding G′ values, indicating a more solid-like behavior as a result of the intermolecular interaction of the dispersion (Li et al.,

**Table 5.2** Sources and the factors influencing rheological characteristics of beta-glucans

| Sources | Factors | Rheological variations | References |
|---|---|---|---|
| High land barley | Molecular weight | Solid-like viscoelastic behavior | (Li et al., 2021) |
| Oat bran | Modification and composition | Higher viscosity | (Kaur et al., 2019) |
| Brewer's yeast | Chemical modification | Higher viscosity | (Kaur et al., 2019) |
| Oats<br>Barley | Molecular size | Enhanced viscosity | (Johansson et al., 2008) |
| Oats | Molecular size | High viscosity with shear thinning behavior | (Mikkelsen et al., 2010) |
| Oats | Concentration, pH, molar mass and temperature | Reduced viscosity | (Hu et al., 2015) |
| *Agaricus bisporus* | Physical modification | Reduced Viscosity | (Ashraf et al., 2015) |

2021). Degenerated β-glucan with lower MW yield in soft gels. This reduced viscosity of beta-glucan is beneficial in beverage commodities since it improves the stability against phase separation while preserving the sensory attributes. Modification techniques that modify the MW and degree of branching of β-glucan facilitate the production of diverse food products with desired viso-rheological qualities (Kaur et al., 2019).

The concentration of the beta-glucan influences the flow behavior of the solutions. The behavior of freshly made oat beta-glucan solutions is identical to that of a neutral random-coil polysaccharide mixture. It acts like a Newtonian fluid with an unaltered viscosity at low concentrations, which indicates the viscosity remains constant irrespective of the shear rate imposed on the solution. The solution exhibits shear-thinning conduct at greater concentrations, where the viscosity lowers with the rising shear rate (Wang & Ellis, 2014). Oat β-glucan demonstrated higher viscosities when contrasted with the barley beta-glucan of the same size. This disparity indicated variations in the fine framework of the oat and barley beta-glucan particularly in the tri and tetraosyl units. The viscosity of oat beta-glucan was found to be cent times greater than barley beta-glucan considering their equivalent beta-glucan concentrations (Johansson et al., 2008). Moreover, the elevated viscosities of the beta-glucan from oats portrayed shear thinning characteristics but barley beta-glucan displayed Newtonian characteristics. The variations in the rheological behavior are ascribed to the difference in structural characteristics involving the molar mass (Mikkelsen et al., 2010). Studies revealed that the viscosity of the oat beta-glucan exhibited a reduction in the viscosity indicating shear thinning non-Newtonian fluid nature which was accelerated as the concentration of the β-glucan raised from 0.1% to 1%. The viscosity of oat beta-glucan displayed a negative correlation with temperature and positively correlated to the molecular weight. The pH of the solution significantly influenced the viscosity where acid and alkaline pH reduced the viscosity in comparison with neutral pH. The viscoelastic behavior of the beta-glucan was afflicted by the temperature, molecular weight, concentration, and pH. Elasticity enhanced and viscosity reduced with the rise in molecular weight and concentration. Conversely, the elasticity and viscosity of the β-glucan dropped with elevated temperatures (Hu et al., 2015).

Physical modification remarkably influences the flow behavior characteristics of the beta-glucan. Irradiation significantly affects the viscosity of the β-D-glucan solution. The β-D-glucan isolated from the *Agaricus bisporus* was treated with 5 to 50 kGy irradiation. The viscosity of the natural β-D-glucan without modification was observed as 191.87 cp. Still, the viscosity dropped to 48.37 cp after the β-D-glucan solution was exposed to a dosage of 50 kGy radiation. The extent of polymerization of *Agaricus* β-D-glucan has declined significantly, which is responsible for this dramatic reduction in the viscosity (Ashraf et al., 2015).

Chemical modification altered the viscous behavior of the beta-glucan. Native β-glucan was discovered to have a greater viscosity than acetylated β-glucan. Beta-glucan will exhibit a higher swelling power due to its better water-retaining capacity between molecules. However, acetylated β-glucan exhibited lessened viscosity, most likely due to a decline in swelling power and less water absorbed between the molecules. The viscosity of the acetylated β-glucan varied when it was subjected to lower shear rates. The viscoelastic properties of natural and acetylated beta-glucan were identical at greater shear rates, indicating that acetylated β-glucan's viscosity remains more stable. In contrast, the viscosity can be modulated by shear rate. In industrial applications, the reduced viscosity brought on by acetylation is advantageous, as it allows for better flow capacity and improved dispersion (De Souza et al., 2015).

## 5.3 Conclusion

Beta-glucans are long-chain polysaccharides made of D-glucose monomers connected by glucosidic connections. The polysaccharide's origin and MW can influence the beta-glucan features. The physicochemical attributes of beta-glucan can be affected by isolation processes, including temperature, pH, particle size, and solvents. Beta-glucan's physical and physicochemical features are crucial in commercial and dietary uses. Beta-glucan utilization as food hydrocolloids has been investigated by understanding their rheological traits. They can form a gel in specific conditions and escalate the product's viscosity. The ability of the beta-glucan to change the texture and expression of food commodities makes them appropriate as thickening agents. The molar ratio and molar mass of the beta-glucan varied with sources and significantly contributed to the rheological behavior. Molecular weight, concentration, temperature, pH, and agronomic conditions remarkably influence the functional attributes of beta-glucan. The physical characteristics of β-glucan, including solubility and rheological conduct in suspensions and gel phases, are modified based on linking patterns, structure, and molecular weight or distributions. These characteristics may be further modified using a variety of techniques and variables, opening up new possibilities for their use in various food, pharmaceutical, and cosmetological applications. In conclusion, β-glucans are flexible substances with potential applications in food industries since they exhibit various physical, functional, and morphological traits. Beta-glucan's distinctive qualities and functioning make it a super ingredient in developing novel functional food and dietary supplements.

## References

Abdel-Haleem, A. M. H., et al. (2020). Characterization of β-glucan gum for food applications as influenced by genotypic variations in three hulless barley varieties. *Journal of Food Science, 85*(6), 1689–1698. https://doi.org/10.1111/1750-3841.15165

Ahmad, A., et al. (2009). Physicochemical and functional properties of barley b-glucan as affected by different extraction procedures. *International Journal of Food Science and Technology, 44*(1), 181–187. https://doi.org/10.1111/j.1365-2621.2008.01721.x

De Arcangelis, E., et al. (2019). Structure analysis of β-glucan in barley and effects of wheat β-glucanase. *Journal of Cereal Science, 85*(December 2018), 175–181. https://doi.org/10.1016/j.jcs.2018.12.002
Ashraf, A., et al. (2015). Effect of γ -irradiation on structural, functional and antioxidant properties of β -glucan extracted from button mushroom (Agaricus bisporus). *Innovative Food Science and Emerging Technologies, 31*, 123–130. https://doi.org/10.1016/j.ifset.2015.05.006
Ashraf, A., et al. (2017). Bioactive Carbohydrates and Dietary Fibre Structural, rheological, antioxidant, and functional properties of β—glucan extracted from edible mushrooms Agaricus bisporus, Pleurotus ostreatus and Coprinus attrimentarius. *Bioactive Carbohydrates and Dietary Fibre, 11*(January), 67–74. https://doi.org/10.1016/j.bcdf.2017.07.006
Bae, I. Y., et al. (2009). Effect of partially hydrolyzed oat β-glucan on the weight gain and lipid profile of mice. *Food Hydrocolloids, 23*(7), 2016–2021. https://doi.org/10.1016/j.foodhyd.2009.03.016
Borchani, C., et al. (2016). Structural characterization, technological functionality, and physiological aspects of fungal β-D-glucans: A review. *Critical Reviews in Food Science and Nutrition, 56*(10), 1746–1752. https://doi.org/10.1080/10408398.2013.854733
Du, B., et al. (2019). A concise review on the molecular structure and function relationship of β-glucan. *International Journal of Molecular Sciences, 20*(16), 4032.
Hamaker, B. R., & Tuncil, Y. E. (2014). A perspective on the complexity of dietary fiber structures and their potential effect on the gut microbiota. *Journal of Molecular Biology, 426*(23), 3838–3850. https://doi.org/10.1016/j.jmb.2014.07.028
Henrion, M., et al. (2019). Cereal B-glucans: The impact of processing and how it affects physiological responses. *Nutrients, 11*(8), 1–14. https://doi.org/10.3390/nu11081729
Herrera, M. P., et al. (2016). β-Glucan content, viscosity, and solubility of Canadian grown oat as influenced by cultivar and growing location. *Canadian Journal of Plant Science, 96*(2), 183–196. https://doi.org/10.1139/cjps-2014-0440
Yuan, H., Lan, P., He, Y., & Chengliang Li, X. M. (2020). Effect of the modifications on the physicochemical and biological properties of β-glucan—A critical review. *Molecules, 25*(1), 57.
Hu, X., et al. (2015). Structure and characteristic of β-glucan in cereal: A review. *Journal of Food Processing and Preservation, 39*(6), 3145–3153. https://doi.org/10.1111/jfpp.12384
Johansson, L., et al. (2008). Comparison of the solution properties of (1→3),(1→4)-β-d-glucans extracted from oats and barley. *LWT, 41*(1), 180–184. https://doi.org/10.1016/j.lwt.2007.01.012
Karp, S., Wyrwisz, J., & Kurek, M. A. (2019). Comparative analysis of the physical properties of o/w emulsions stabilised by cereal β-glucan and other stabilisers. *International Journal of Biological Macromolecules, 132*, 236–243. https://doi.org/10.1016/j.ijbiomac.2019.03.212
Kaur, R., Sharma, M., Ji, D., Xu, M., & Agyei, D. (2019). Structural features, modification, and functionalities of beta-glucan. *Fibers, 8*(1), 1–29.
Khan, A. A., et al. (2016). Structural, thermal, functional, antioxidant & antimicrobial properties of β-d-glucan extracted from baker's yeast (Saccharomyces cereviseae) – Effect of γ-irradiation. *Carbohydrate Polymers, 140*, 442–450. https://doi.org/10.1016/j.carbpol.2016.01.003
Lante, A., Canazza, E., & Tessari, P. (2023). Beta-glucans of cereals: Functional and technological properties. *Nutrients, 15*(9), 2124.
Lazaridou, A., Biliaderis, C. G., & Izydorczyk, M. S. (2007). Cereal beta-glucans: structures, physical properties, and physiological functions. In *Functional food carbohydrates* (pp. 1–72).
Lazaridou, A., & Biliaderis, C. G. (2007). Molecular aspects of cereal β-glucan functionality: Physical properties, technological applications and physiological effects. *Journal of Cereal Science, 46*(2), 101–118. https://doi.org/10.1016/j.jcs.2007.05.003
Li, W., et al. (2011). Studies of aggregation behaviours of cereal b -glucans in dilute aqueous solutions by light scattering: Part I. Structure effects. *Food Hydrocolloids, 25*(2), 189–195. https://doi.org/10.1016/j.foodhyd.2010.02.005
Li, Y., et al. (2021). Comparison of distribution and physicochemical properties of β-glucan extracted from different fractions of highland barley grains. *International Journal of Biological Macromolecules, 189*(2), 91–99. https://doi.org/10.1016/j.ijbiomac.2021.08.094

Maheshwari, G., Sowrirajan, S., & Joseph, B. (2017). Extraction and isolation of β-glucan from grain sources—A review. *Journal of Food Science, 82*(7), 1535–1545. https://doi.org/10.1111/1750-3841.13765

Mejía, S. M. V., de Francisco, A., & Bohrer, B. M. (2020). A comprehensive review on cereal β-glucan: Extraction, characterization, causes of degradation, and food application. *Critical Reviews in Food Science and Nutrition, 60*(21), 3693–3704. https://doi.org/10.1080/10408398.2019.1706444

Mikkelsen, M. S., et al. (2010). Comparative spectroscopic and rheological studies on crude and purified soluble barley and oat β-glucan preparations. *Food Research International, 43*(10), 2417–2424. https://doi.org/10.1016/j.foodres.2010.09.016

Mikkelsen, M. S., et al. (2013). Molecular structure of large-scale extracted β-glucan from barley and oat: Identification of a significantly changed block structure in a high β-glucan barley mutant. *Food Chemistry, 136*(1), 130–138. https://doi.org/10.1016/j.foodchem.2012.07.097

Nakashima, A., et al. (2018). Β-glucan in foods and its physiological functions. *Journal of Nutritional Science and Vitaminology, 64*(1), 8–17. https://doi.org/10.3177/jnsv.64.8

Oliveira, L. d. C., et al. (2012). Avaliação da temperatura de secagem na qualidade de β-glicanas em grãos de aveia branca. *Ciência e Tecnologia de Alimentos, 32*(4), 793–797. https://doi.org/10.1590/S0101-20612012005000105

Panel, E., & Nda, A. (2011). Scientific opinion on the substantiation of health claims related to beta-glucans from oats and barley and maintenance of normal blood LDL-cholesterol concentrations (ID 1236, 1299), increase in satiety leading to a reduction in energy intake (ID 851, 852). *EFSA Journal, 9*(6). https://doi.org/10.2903/j.efsa.2011.2207

Qin, Y., et al. (2021). Effect of acid and oxidative degradation on the structural, rheological, and physiological properties of oat β-glucan. *Food Hydrocolloids, 112*(May 2020), 106284. https://doi.org/10.1016/j.foodhyd.2020.106284

Regand, A., et al. (2011). The molecular weight, solubility and viscosity of oat beta-glucan affect human glycemic response by modifying starch digestibility. *Food Chemistry, 129*(2), 297–304. https://doi.org/10.1016/j.foodchem.2011.04.053

Ryu, J. H., et al. (2012). Effects of barley and oat β-glucan structures on their rheological and thermal characteristics. *Carbohydrate Polymers, 89*(4), 1238–1243. https://doi.org/10.1016/j.carbpol.2012.04.025

Schmidt, M. (2022). Cereal beta-glucans: An underutilized health endorsing food ingredient. *Critical Reviews in Food Science and Nutrition, 62*(12), 3281–3300. https://doi.org/10.1080/10408398.2020.1864619

Sharma, S., Saxena, D. C., & Riar, C. S. (2018). Characteristics of β-glucan extracted from raw and germinated foxtail (Setaria italica) and kodo (Paspalum scrobiculatum) millets. *International Journal of Biological Macromolecules, 118*, 141–148. https://doi.org/10.1016/j.ijbiomac.2018.06.064

Skendi, A., et al. (2003). Structure and rheological properties of water soluble b -glucans from oat cultivars of Avena sativa and Avena bysantina. *Journal of Cereal Science, 38*, 15–31. https://doi.org/10.1016/S0733-5210(02)00137-6

De Souza, N. L., et al. (2015). Functional, thermal and rheological properties of oat β-glucan modified by acetylation. *Food Chemistry, 178*, 243–250. https://doi.org/10.1016/j.foodchem.2015.01.079

Singh, S., & Maninder Kaur, D. S. S. (2018). β-glucan from different Indian oat (Avena sativa) cultivars: Chemical, functional, structural, and rheological properties. *Journal of Food Processing and Preservation, 42*(11), e13788.

Sun, T., et al. (2020). Rheological and functional properties of oat β-glucan with different molecular weight. *Journal of Molecular Structure, 1209*, 127944. https://doi.org/10.1016/j.molstruc.2020.127944

Tosh, S. M., et al. (2004). Structural characteristics and rheological properties of partially hydrolyzed oat b -glucan: the effects of molecular weight and hydrolysis method. *Carbohydrate Polymers, 55*, 425–436. https://doi.org/10.1016/j.carbpol.2003.11.004

Vaikousi, H., Biliaderis, C. G., & Izydorczyk, M. S. (2004). Solution flow behavior and gelling properties of water-soluble barley (1→3, 1→4 ) – b -glucans varying in molecular size. *Journal of Cereal Science, 39*, 119–137. https://doi.org/10.1016/j.jcs.2003.09.001

Wang, Q., & Ellis, P. R. (2014). Oat β-glucan: Physico-chemical characteristics in relation to its blood-glucose and cholesterol-lowering properties. *British Journal of Nutrition, 112*, S4–S13. https://doi.org/10.1017/S0007114514002256

Whitehead, A., et al. (2014). Cholesterol-lowering effects of oat β-glucan: A meta-analysis of randomized controlled trials1. *American Journal of Clinical Nutrition, 100*(6), 1413–1421. https://doi.org/10.3945/ajcn.114.086108

Zhang, M., Bai, X., & Zhang, Z. (2011). Extrusion process improves the functionality of soluble dietary fiber in oat bran. *Journal of Cereal Science, 54*(1), 98–103. https://doi.org/10.1016/j.jcs.2011.04.001

Zhang, Y., et al. (2023). Recent advances of cereal β -glucan on immunity with gut microbiota regulation functions and its intelligent gelling application. *Critical Reviews in Food Science and Nutrition, 63*(19), 3895–3911. https://doi.org/10.1080/10408398.2021.1995842

Zhu, F., Du, B., & Xu, B. (2016). A critical review on production and industrial applications of beta-glucans. *Food Hydrocolloids, 52*, 275–288. https://doi.org/10.1016/j.foodhyd.2015.07.003

# Chapter 6
# Modifications of Beta-Glucan

**Muhammed Navaf and Kappat Valiyapeediyekkal Sunooj**

## 6.1 Introduction

Polysaccharides are a vital component in many foods. Polysaccharides may be digestible or indigestible and function as a structural material and energy source. Digestion of polysaccharides provides energy to the body that needs to function normally. There have also been studies showing that a variety of indigestible polysaccharides are beneficial to the body (Ge et al., 2021). β-glucans are polysaccharides with known bioactive functionalities and consist of D-glucose units linked with β-(1–3), β-(1–4), and/or β-(1–6) glycosidic linkages (Fig. 6.1) either in a linear or branched manner (Kaur et al., 2020). β-glucan activates the macrophage to initiate anti-tumor, antibacterial, wound healing, anti-oxidant, etc., (Yuan et al., 2020). Additionally, it reduces glycemic index and serum cholesterol; controls diabetes, hypertension, etc. (Kaur et al., 2020). β-glucan have the potential to form viscous gels even at low concentration and can be used as a thickening agent (De Souza et al., 2015).

β-Glucans are found in various natural sources and found in larger quantities in seeds of some cereals (oats, barley) and the cell walls of higher plants. It can also be obtained from mushrooms, yeasts, seaweed, algae, and molds.

The potential biological activity of various β-glucans is given below

- Cereal β-D-glucan exhibits prominent cholesterol-lowering activity.
- Mushroom β-glucan exhibits potential anticancer properties.
- Yeast β-glucan exhibits immune-enhancing properties.
- Seaweed β-glucan exhibits anti-tumor and immunomodulatory effects.
- Algae β-glucan exhibits anti-oxidant, anti-inflammatory, and immune-stimulating properties.

M. Navaf · K. V. Sunooj (✉)
Department of Food Science and Technology, Pondicherry University, Puducherry, India



M. Bashir Mir, S. A. Mir (eds.), *Beta-Glucan: Sources, Properties and Applications*,
https://doi.org/10.1007/978-3-031-95788-8_6

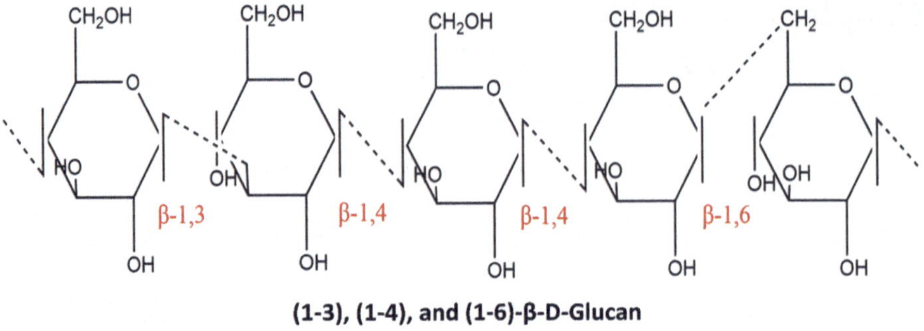


**Fig. 6.1** β-glucan structure

Cereal β-glucans are made up of β-(1–3) and β-(1–4) glycosidic linkages, besides yeast, fungi, and mushroom β-glucans are made up of β-(1–3) and β-(1–6) glycosidic linkages. It can be found either in a random coil or an organized conformation with single helix chains associated as triple helices that are stabilized by inter and intra-molecular hydrogen bonds. The intramolecular polyhydroxy groups of β-glucans interact with each other forming a dense triple helical structure, and it is partially soluble or insoluble in nature (Yuan et al., 2020). The solubility of β-glucan is strongly influenced by its molecular weight (MW), branching pattern, and spatial arrangement. Generally, low MW β-glucan (less than 400 kDa) has better water solubility (Wu et al., 2021). However, the high MW branched β-glucans are insoluble in nature. Among their various applications, such as food additives, β-glucans solubility influences their rheological, nutritional, and sensory properties (Kaur et al., 2020).

The high MW β-glucans are insoluble in nature; hence it is unsuitable for many food and medical applications in their intact form. For example, Yeast β-glucans have high MW, making it difficult for them to pass through cell membranes and perform their functional properties. In order to enhance solubility and biological activity, β-glucans need to be modified. All active β-glucan possess similar structures and their modification results in conformational changes that will alter the solubility and bioactivity (Yuan et al., 2020). The common techniques used to modify β-glucan include physical, chemical, enzymatic, and mechanical modifications (Fig. 6.2).

## 6.2 Physical Modifications

The physical modifications of β-glucan enhance the solubility and alter the functional properties. Physical modifications result in the degradation or breakdown of the glycosidic bond of the β-glucan chain without destroying its basic structure. The physical modifications of β-glucan can be achieved by different methods, i,e, thermal degradation, irradiation, high-pressure treatment, and microwave treatment.

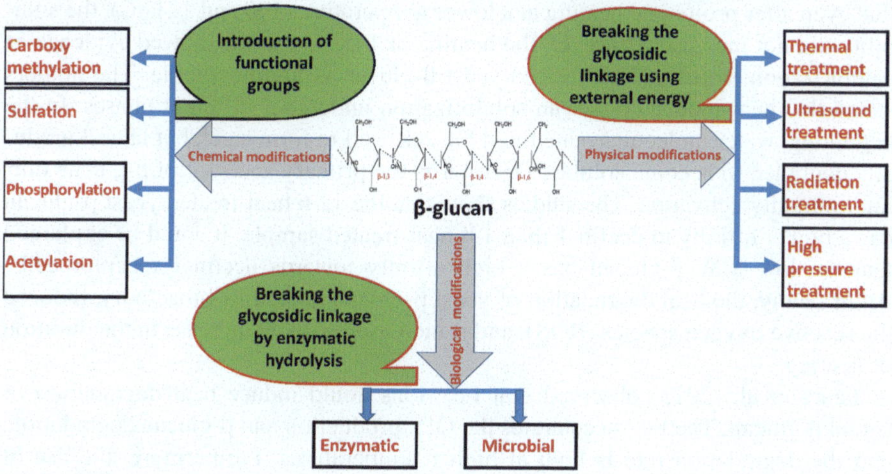


**Fig. 6.2** Major methods of β-glucan modification

Physical modifications of β-glucan lead to MW reduction as well as increased solubility, improved functional properties, and conformational changes in the solution.

## 6.2.1 *Thermal Degradation*

Thermal degradation is one of the most important methods of modification of polymers. It is a simple method, and uniform products can be produced easily. When heat is applied with low energy, structural breakdown occurs, causing the side chains to rotate in a triple helix configuration (Xu et al., 2013). The heating of β-glucan at 150 °C in a solution efficiently decreases the molecular weight (MW) without altering its primary structure. This reduction in MW may alter the ordered arrangement of β-glucan to a randomly coiled structure (Adachi et al., 1990). Xu et al. (2013) observed that heating of jelly ear fungus β-glucan at 25–150 °C decreased the MW from $3.50 \times 10^6$ to $2.67 \times 10^6$. Similarly, heating at 25–150 °C is also decreased the molecular size from 185 to 160 nm. The heating of β-glucan breaks down the inter and intramolecular hydrogen bonds, resulting in lower MW β-glucan formation. Thus reduction in MW alters the functionalities of β-glucan. It includes the reduction in viscosity, hydrodynamic radius, radius of gyration, etc. In addition, structural change from a rod-like chain to a flexible configuration was observed at a temperature of 140 to 160 °C.

The yeast β-glucans have an immunomodulatory effect and are insoluble in their native form. Hence, the solubilization of yeast β-glucan will intensify its utility in the food and medical fields. Ishimoto et al. (2017) heated yeast β-glucan at different temperatures, i.e., 100, 121, and 135 °C for different times (0 to 24 h), and observed that even after prolonged heating at a lower temperature (100 and 121 °C), the solubility did not increase. However, the heating at 135 °C for 1 h showed evidence of solubilization, and the 24 h treated yeast β-glucan is highly soluble. The authors stated that heat-induced β-glucan solubilization undergoes different phases. In the beginning, water molecules penetrated the cell wall to form a gel, but later it gradually degraded to become soluble. Dectin-1 is the primary acceptor of β-glucan during immunity activation. The studies show that the 12 h heat-treated yeast β-glucan has a higher affinity to dectin-1 than 1 h heat-treated sample. It could be explained that the low MW β-glucan has a high affinity towards dectin-1 receptor cells. Additionally, the heat degradation of yeast β-glucan altered its bioactivity, reduced the reactive oxygen species (ROS), and it increased with an increase in the duration of heating.

Faure et al. (2013) observed that $Fe^{2+}$ ions could induce heat degradation of cereal β-glucan. The $Fe^{2+}$ accelerates the $OH^*$ production and β-glucan degradation, and the degradation rate is high at higher temperatures. Furthermore, the use of ascorbic acid along with $Fe^{2+}$ causes degradation at a lower temperature (25 °C). MW and concentration are crucial factors that affect β-glucan viscosity. Thus, the decrease in MW can explain the reduction in viscosity. The heating of cereal β-glucan at 25, 50, and 100 °C for 24 h with 50 μM $Fe^{2+}$ decreased the viscosity by 7, 16, and 91%, respectively. Furthermore, the addition of a combination of ascorbic acid and $Fe^{2+}$ resulted in a significantly higher (72%) reduction in β-glucan viscosity.

### *6.2.2 Irradiation*

Irradiation is a physical treatment that alters the physicochemical characteristics of polysaccharides by exposing them to ionizing radiations, i.e., gamma radiation, X-ray, electron beam, etc. Radiation-induced degradation of polymers results in reproducible and quantitative changes without using chemicals and temperature control devices (Byun et al., 2008). The irradiation of β-glucan causes the breakdown and degradation of the structure, resulting in a potential molecular weight (MW) reduction, and the extent of breakdown increases with an increase in radiation dose. The radiolysis of glycosidic linkages of β-glucan changes the basic structure, and fragmented polysaccharides are produced. It gives an improved solubility and decreased viscosity to the irradiated β-glucan (Yuan et al., 2020). In addition, it facilitates the availability of small β-glucan into targets, thus helping in enhanced functional properties like anti-oxidant activity, antiproliferative activity, etc. (Shah et al., 2015b).

Byun et al. (2008) observed that gamma irradiation of black yeast (*Aureobasidium* sp) β-glucan with 10–50 kGy significantly decreased the average MW with an increase in radiation dose. The native β-glucan has a MW of 178 kDa and was decreased to 62, 32, and 25 kDa after the irradiation of 10, 30, and 50 KGy, respectively. In addition, enhanced solubility and decreased viscosity were observed. A 73% reduction in viscosity was observed after 50 kGy of radiation. The morphological characteristics of irradiated β-glucan showed that the irradiation resulted in the distortion of β-glucan into smaller granules. Sung et al. (2009) reported enhanced macrophage cell (RAW 264.7) stimulation activities, increased cell viability, and cytokine release (IFN-g and IL-2) of spleen and Peyer's patch cells in irradiated black yeast β-glucan. It was also observed that macrophage cell proliferation is dose-dependent.

The gamma irradiation of barley (*Hordeum vulgare*) β-glucan with a dose of 8 kGy reduced the MW from 177 kDa to 79 kDa (Shah et al., 2015a). Furthermore, the FTIR analysis showed the formation of new functional groups, i.e., -COOH, -OH, and C=O in irradiated β-glucan, which enhanced the radical scavenging activity and in turn increased the antioxidant activity. The irradiation of barley β-glucan with a dose of 8 kGy showed a significant increase in reducing power, whereas a lower dose (2 and 4 kGy) did not show any difference from untreated β-glucan. Similarly, a dose-dependent increase in chelating ability was observed, and it increased from 7% to 52% by irradiation of 8 kGy. The irradiation with 4 and 8 kGy increased the inhibitory effect of lipid oxidation by 13% and 18%. The lipid oxidation is mainly due to the OH* generation. Thus, the reduction in lipid oxidation of β-glucan by irradiation could be ascribed to the presence of carbonyl groups in irradiated β-glucan. It interacts with transition metals that initiate oxidation. Further, the irradiation of β-glucan also increases the antiproliferative activity. A 37% increase in antiproliferative activity was observed in 8 kGy irradiated barley β-glucan.

Shah et al. (2015b) noted that a 10 kGy irradiation of oat β-glucan reduced the average MW from 199 kDa to 70 kDa. Similarly, its viscosity was reduced from 59.99 to 23.33 cP with irradiation of 10 kGy. A low dose (<10 kGy) irradiation of oat β-glucan did not cause any noticeable effect in reducing power and lipid oxidation inhibition capacity, whereas it increased by 10 kGy irradiation. It also showed a dose-dependent increase in antioxidant and antiproliferative activity. A 6 and 10 kGy irradiation of oat β-glucan increased the antiproliferative activity against colo-205 cancer cells to 22% and 40%, respectively. Further, 10 kGy irradiation leads to 20% and 24% antiproliferative activity against T47D tumor and MCF7 cell lines, respectively.

Similarly, the irradiation of yeast β-glucan by 50 kGy decreased MW from 175 kDa to 27.9 kDa. The thermal analysis shows that the irradiation of yeast β-glucan increased the onset and peak temperature and enthalpy of gelatinization. The irradiation brings structural changes in yeast β-glucan, leading to a lower water retention capacity, resulting in decreased swelling power and viscosity. A dose-dependent increase in emulsion capacity and stability, foaming capacity and stability, and fat, and bile acid binding capacity were noticed. In addition, it also exhibited improved antioxidant activity and chelating ability. The degradation of β-glucan by irradiation lead to the formation of low MW β-glucan, which can penetrate the cell wall of microbes and affects cell metabolism, leading to the death of microbes. The antimicrobial assay of irradiated yeast β-glucan shows that it has antibacterial activity against bacteria such as *E. coli, S. aureus, B. subtillis, and S. typhi* (Khan et al., 2016).

### 6.2.3 Ultrasonication

This method uses low-frequency, high-intensity ultrasound energy to increase the molecular vibration of particles and break some of the pyran dextran chemical bonds. Polysaccharides can be efficiently and effectively modified by ultrasonication. During ultrasonication, it depolymerizes polysaccharide molecules and breaks their side chains, resulting in the formation of an oligosaccharide molecule (Yuan et al., 2020). The ultrasonication of *Leuconostoc citreum* β-glucan for 5 and 160 min gradually shifted the molecular weight (MW) to a lower MW region with a narrow distribution. This shift could be due to the mechanical degradation of high MW β-glucan to low MW β-glucan.

The ultrasonication of *Poria cocos* β-glucan in an aqueous medium changed its compact structure into an open-chain configuration. A 10 and 90 min sonication reduced the MW from 251 to 104 kDa and 251 to 43 kDa, respectively. Similarly, 10 and 90 min sonication increased the solubility from 2.48 to 3.31 mg/mL and 2.48 to 6.30 mg/mL, respectively (Chen et al., 2015). The increased water solubility of ultrasound-treated β-glucan could be due to the degradation of the large MW polymer chain and an associated increase in short-chain polymers; this will enhance the hydration tendency of β-glucan.

Short-term ultrasound treatment of *Lignosus rhinocerotis* β-glucan changes the color from milky white to transparent for a sonication time of up to 10 min. However, a further increase in sonication time (>30 min) caused the development of turbidity. The solubility of ultrasound-treated *Lignosus rhinocerotis* β-glucan also increased after 10 min treatment; however, a further increase in treatment time decreased the solubility. The particle size analysis showed that in a short treatment time, the particle size reduced from 964 to 110 nm, whereas prolonged treatment increased the particle size to 5085.67 nm. It was stated that the heating effect is more dominant during prolonged sonication than the cavitation effect, resulting in the re-agglomeration of smaller particles and increasing MW (Hu et al., 2021). It is possible to destroy glycosidic bonds using high-frequency ultrasound, but sugar rings and triple helices cannot be destroyed by high-frequency ultrasound. A short period of ultrasonic cavitation can result in the disaggregation of spherical glucan aggregates with hyperbranched chains into triple helices; they can be reaggregated with long-term ultrasonic treatment.

Yuan et al. (2022) noticed that ultrasound-assisted-enzymatically extracted yeast β-glucan showed higher solubility at a lower ultrasonic intensity and lower solubility at higher ultrasonic intensity. Maximum solubility was noticed at an intensity of 18 W/mL. Likewise, the solubility of β-glucan hydrolyzed for 10 min by a single enzyme increased from 11.65% to 70.42%. The DPPH (2,2-diphenyl-1-picrylhydrazyl) scavenging activity was also increased and is dependent on the enzyme concentration. The thermal analysis of ultrasonicated yeast β-glucan showed that the degradation temperature decreased from 30–230 °C to 30–170 °C. Likewise, the thermal transition temperature dropped from 118 to 95 °C after sonication. The ultrasound-assisted enzymolysis of yeast β-glucan bring down the particle size from 8.80 to 1.77 μm (Zheng et al., 2019).

### 6.2.4 High-Pressure (HP) Treatment

An innovative approach to food processing is the use of dynamic high pressure (HP) microjets, which combine HP with homogenization technology. In these devices, ultrasonic radiation and mechanical shear are used to transfer mechanical energy to fluid particles under HP, resulting in changes in their structure and molecular weight (MW) (Yuan et al., 2020). It modifies the materials in the oscillating reaction chamber through high-speed collisions, high-frequency oscillations, instantaneous pressure drops, strong shear, cavitation, and other methods (Benchamas et al., 2020).

The HP modification of β-glucan decreases the MW, alters the microstructure, and enhances hygroscopicity and solubility. Liu et al. (2016) synthesized water-soluble yeast β-glucan using HP, and the NMR spectrum showed that the treatment destroyed the hydrogen bond. The solubility of HP treated β-glucan depends on the pressure applied, treatment time, and the ratio of β-glucan to the solvent.

### 6.2.5 Microwave Modification

Microwave modification of β-glucan causes molecular chain stretching; it may alter its physical and chemical characteristics (Q. Wang et al., 2017). Tao and Xu (2008) reported a complete solubilization of king tuber mushroom β-glucan in 0.02% (w/w) $NaN_3$ solution when it is treated with microwave radiation with an optimum microwave condition of 35 s at 765 W. They also noticed structural changes in microwave-treated β-glucan. At 0.02% (w/w) $NaN_3$ solution, the microwave-treated β-glucan existed in a sphere-like conformation.

### 6.2.6 Other Physical Modification Methods

The quality of cereal grains may be impacted by post-harvest drying. In a study by Zechner-Krpan et al. (2010), different drying methods were investigated for spent brewer's yeast β-glucan. It was observed that agglomeration and microstructural changes were caused by air drying and lyophilization. However, the agglomerations and microstructure were not significantly altered by spray drying.

Oliveira et al. (2010) observed the changes in water holding capacity (WHC) and gelation capacity of oat β-glucan after drying at different temperatures (25, 50, 75, and 100 °C). It was observed that the drying at a lower temperature (≤50 °C) did not cause noticeable changes in WHC. At the same time, drying at higher temperatures (75 and 100 °C) reduced the WHC. In addition, the high-temperature drying adversely affected the β-glucan structure and altered the gel-forming capacity. Fu et al. (2022) reported that freeze-drying significantly impacts yeast β-glucan. They observed that freeze-dried β-glucan showed high WHC, oil binding, and swelling capacity as compared to hot air and vacuum-dried β-glucan. It was also observed that the freeze-drying significantly improved the bile acid binding ability of yeast β-glucan. Besides, it inhibited the activity of enzymes like pancreatic lipase, α-amylase, and α-amyloglucosidase.

## 6.3 Chemical Modifications

The chemical modifications of β-glucan primarily depend on the number of hydroxyl groups (–OH) on the β-glucan chain, since the –OH is mainly involved in the reaction of β-glucan with a modifying agent. The addition of functional groups to β-glucan enhances its specific functionality. The chemically modified β-glucan has tremendous application in food and pharmaceuticals. The β-glucan can be chemically modified by different methods, like carboxymethylation, sulfation, phosphorylation, acetylation, oxidation, and acid hydrolysis.

### *6.3.1 Caroboxymethylation*

Carboxymethylation is a widely used method for β-glucan modifications and mainly aims to enhance its water solubility. Carboxymethylation can be done by replacing the –OH of polysaccharides with the $–CH_2–COOH$ (Fig. 6.3). It alters the hydrophilicity of β-glucan, thereby increasing the water solubility and leading to improved functional properties (Yongjiang Wang et al., 2009). Carboxymethylated β-glucan has tremendous applications in different industries, such as chemical, pharmaceutical, food, cosmetic, etc. (Francini Yumi Kagimura et al., 2015b). Carboxymethylated β-glucan exhibits bioactivity by reducing sterol levels. Carboxymethylated polysaccharides are generally synthesized by reacting polysaccharides in alkaline

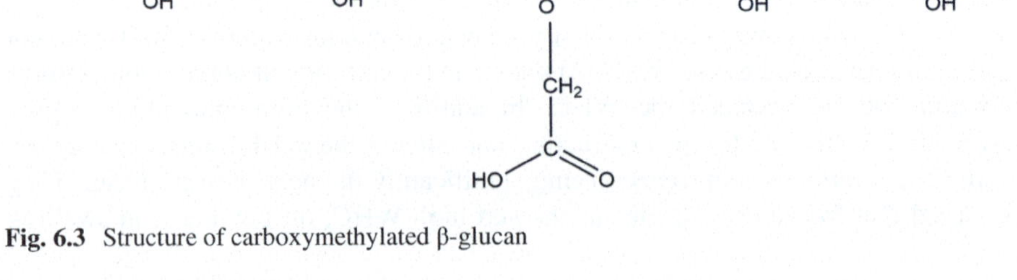


**Fig. 6.3** Structure of carboxymethylated β-glucan

2-propanone with chloroacetic acid or sodium mono-chloroacetate. Carboxymethylation has the advantage of having reactive carboxymethyl groups on the polysaccharides that are distant from the polysaccharide backbone and are readily involved in reactions (Taubner et al., 2020). Carboxymethylation enhances the β-glucan solubility and alters the original conformation of the dextran molecule and its physiological activities. The carboxymethylated β-glucan changes its structure from a triple-helical to a single-helical and then to a random structure as the degree of substitution (DS) increases (Yuan et al., 2020).

The carboxymethylation of barley bran β-glucan enhances its antimicrobial activity (Song et al., 2020). Carboxymethylated β-glucan altered the membrane permeability and causes morphological and structural damage to *Staphylococcus aureus* cells, resulting in the degradation of the cell wall and membrane and, finally, the cell lysis. Ohno et al. (1988) observed an enhanced anti-tumor activity of carboxymethylated β-glucan. They reported the effect of varying DS (0.04–0.49) of carboxymethyl *Sclerotinia sclerotiorum* β-glucan. It was observed that carboxymethylated β-glucan with a DS 0.04–0.09 potentially decreased the viscosity. In addition, the anti-tumor activity varied with the DS, and a DS of 0.14 showed prominent anti-tumor activity.

Fujiike et al. (2022) synthesized carboxymethylated botryosphaeran, a fungal β-glucan, and found it to be more water-soluble. In addition, it exhibited excellent anticancer activity against MCF-7 cells. A 24 h incubation restricted the proliferation of MCF-7 cells. Likewise, the carboxymethylated yeast cell wall β-glucan acts as a potent antioxidant. It shows free radical scavenging activity even at a low concentration and can be comparable to other known anti-oxidants (Babincová et al., 2002).

Yifeng Wang and Zhang (2006) reported that, in 0.2 M NaOH solution, the carboxymethylation of *Poria cocos* β-glucan makes the β-glucan highly flexible. It also increased the β-glucan MW from 3440 to 3990 kDa as the DS increased from 0.26 to 0.91 (Yongjiang Wang et al., 2009). The researchers also observed an increased solubility and bile acid binding capacity of the carboxymethylated *Poria cocos* β-glucan. The hydrophilic characteristics of the carboxymethyl group facilitate the enhanced solubility of carboxymethyl derivatives, and the higher bile acid binding capacity explains its possible effect on lowering body cholesterol. Carboxymethyl β-glucan hastens intestinal viscosity and reduces the cholesterol and bile acid absorption in the body. As compared to native *Poria cocos* β-glucan, the carboxymethyl derivative showed an enhanced antioxidant activity (DPPH scavenging activity). Similarly, carboxymethylation of *Poria cocos* β-glucan enhanced its antitumor activities in sarcoma 180 tumor cells. At a 20 × 8 mg/kg/day dose, administration of carboxymethyl derivative increased the inhibition ratio from 2.5% to 41.8% (X. Chen et al., 2010).

The carboxymethylated *Pleurotus tuber-regium* sclerotia β-glucan shows antitumor activity in HL-60 and HepG2 cell lines. An increase in carboxymethyl β-glucan content from 50 to 150 μg/mL increased the inhibition ratio from 28.44% to 43.21%, respectively (M. Zhang et al., 2004a, b). Several factors affect the anti-tumor activity of β-glucans, including sugar linkages, branching length, DS, molecular size, and conformation (M. Zhang et al., 2003).

The carboxymethylation of *Lasiodilpodia theobromae* β-glucan enhanced the water solubility from 3% to 69%. Similarly, the carboxymethyl derivative of *Lasiodilpodia theobromae* β-glucan showed enhanced antioxidant activity. It could be due to the increased electron density around the molecule, resulting in more active hydroxyl group formation and increasing the hydrogen donation tendency (Francini Y. Kagimura et al., 2015a). The carboxymethylation of qingke β-glucan significantly enhances its antioxidant activity. The addition of the carboxymethyl group to qingke β-glucan enhanced the NO (nitric oxide) and DPPH radical scavenging activity. Similarly, carboxymethyl qingke β-glucan exhibited an increased *in-vitro* binding ability for fat, bile acid, and cholesterol and inhibition activity toward pancreatic lipases (H. Guo et al., 2020).

### 6.3.2 Sulfation

Sulfation is another method of β-glucan modification and it is a widely practiced method for polysaccharide modification. The synthesis of sulfated β-glucan involves replacing –OH of polysaccharides with sulfate groups (Fig. 6.4), which improves their biological activity. Sulfated β-glucans have several advantages, including high yield, high solubility in water, anticoagulant, antithrombotic, and anticancer properties, and promote inflammation and immune responses (C. Chen et al., 2014). In particular, the sulfate group's hydrophilic characteristics increase the modified β-glucan solubility. Polysaccharides' polarity and aqueous solubility can be improved by novel glucan sulphation. The improved aqueous solubility of sulfated polysaccharides greatly benefits its application, particularly in pharmaceuticals. The DS can be used to evaluate the efficacy of sulfate modification. According to the consensus, polysaccharides bind more sulfate groups, when they have higher DS.

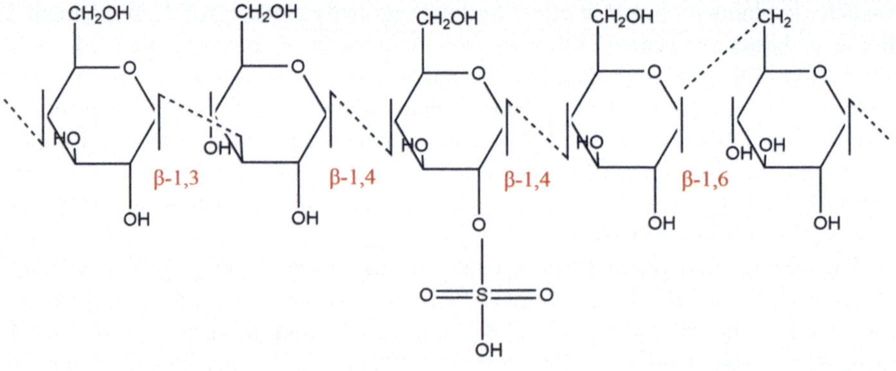


**Fig. 6.4** Structure of sulfated β-glucan

Zhang et al. (2021) reported the impact of different reaction times on the DS and the MW of oat β-glucans. They observed that as the reaction duration increased, the DS and MW increased. Moreover, the sulfation significantly increased the apparent viscosity and altered the viscoelastic characteristics of β-glucans. The sulfation of acid-degraded β-glucan increased the fat and bile acid binding capacity, and it increased with the DS and MW. Similarly, the sulfation of oxidatively degraded β-glucans also enhanced the fat and bile acid binding capacity. However, contrary to the above, sulfation of oxidatively degraded β-glucan shows a decrease in fat and bile binding ability with an increase in DS and MW. In addition, the sulfation of both acid-degraded and oxidatively degraded β-glucans showed a decreased α-glucosidase activity and glucose availability. Guo et al. (2019) modified the qingke (Tibetan hulless barley) β-glucans using sulfur trioxide-pyridine. They observed that the sulfation significantly changed the water solubility, apparent viscosity, MW, and molar ratio. In addition, sulfated qingke showed enhanced antioxidant activities and in-vitro binding properties and decreased pancreatic lipase inhibition activity.

Bae et al. (2008) modified barley β-glucans using formamide and chlorosulfonic acid and studied its anticoagulant activity. It was observed that the sulfated barley β-glucans exhibited anticoagulant activity. In addition, it prolonged the blood clotting time, and it is dependent on concentration. Furthermore, the sulfate group content and the MW also depend on its anticoagulant activity. The anticoagulant activity of sulfated β-glucan is ascribed to the anionic interaction of sulfated β-glucan with the positive charge of the proteins involved in coagulation.

Yucel Falco et al. (2017) also used formamide and chlorosulfonic acid for the synthesis of sulfated oat β-glucan. The researchers studied the suitability of the modified β-glucan for the synthesis of protective coating for targeted delivery to the gastrointestinal system. They noted that sulfated β-glucan helps in the synthesis of multilayered prebiotic coating. It decreased the cell count by 2 log cfu/mL, showing resistance to an acidic environment in the gastric system. Thus sulfated β-glucan has the potential application as a carrier for prebiotic and delicate nutraceuticals.

The sulfated β-glucans have potential pharmacological activities. The sulfation of β-glucans extracted from different sources, i.e., Poria mushroom and king tuber oyster mushroom, have anti-viral and anti-tumor activities. Similarly, sulfated *Parmotrema* mushroom β-glucans exert antithrombotic and anticoagulant effects. Baranoski et al. (2015) studied the effect of sulfated *Agaricus brasiliensis* β-glucans on MCF-7 cell line. They observed that at 5 mg/mL concentration of sulfated β-glucan exhibited cytotoxicity, and the concentration of sulfated β-glucan increased to 100 mg/mL, proliferation was decreased after 72 h.

The β-glucan extracted from *poria cocos* have anti-tumor activity against the S-180 tumor cells when carboxymethylated and sulfated. It also enhanced the immune response of the spleen. In addition, it caused necrosis and apoptosis of tumor cells. The immune enhancement test of carboxymethylated and sulfated β-glucan displayed an increase in the carbon clearance index of macrophage, spleen antibody production, etc., in mice (X. Chen et al., 2010). Similarly, the *in-vitro* assay of sulfated *saccharomyces cerevisiae* β-glucan on the splenic lymphocyte showed that the sulfated β-glucan significantly enhanced the lymphocyte proliferation alone or in combination with concanavalin A and lipopolysaccharides (LPS) and stimulated the cell to secrete IL-2 and INF-γ (M. Wang et al., 2016).

The *in-vitro* and *in-vivo* studies of sulfated *Phellinus ribis* β-glucan showed an improved anti-tumor (H22 hepatocellular carcinoma) activity without producing any sign of toxicity (Y. Liu et al., 2014). The immunohistochemistry assay showed that, in tumor cells, β-glucan potentially decreased microvessel density and inhibited the vascular endothelial growth factor expression. Similarly, sulfated fungi (*Pleurotus tuber-regium*) β-glucans with a definite MW exhibited anti-viral activity against HSV-1) and HSV-2 cell lines (M. Zhang et al., 2004a, b). Sousa et al. (2022) reported that the *in-vitro* analysis of sulfated *Agaricus subrufescens* β-glucans exhibited efficacy against dengue and zika viruses as well as the nonstructured protein of these virus-induced epithelial damages. Likewise, a $5.98 \times 10^{-4}$ $\mu$M (EC50) sulfated derivative of oat β-glucans exhibited potent anti-viral activity against HIV-1 cells (Shun et al., 2008).

## *6.3.3 Phosphorylation*

Phosphorylation modification of polysaccharides introduces a phosphate group into polysaccharides. During phosphorylation, the hydroxyl group polysaccharide chain is replaced with a phosphate group (Fig. 6.5). Phosphorylation enhances the solubility and bioactivities of β-glucans. In comparison to unmodified β-glucans, phosphorylated β-glucans show potent anti-inflammatory, anti-viral, and antiproliferative properties (Francini Yumi Kagimura et al., 2015b). In general, phosphorus oxychloride, phosphoric anhydride, phosphoric acid, or their salts are used as phosphorylating agents in β-glucan modification. Strong acids are needed as a catalyst for phosphorylation, which often results in complex polysaccharides being degraded (Yuan et al., 2020). In 1990, Williams et al., proposed β-glucans modification by phosphorylation. They phosphorylated the *saccharomyces cerevisiae* β-glucans using dimethyl sulfoxide. The developed β-glucan phosphate possesses excellent

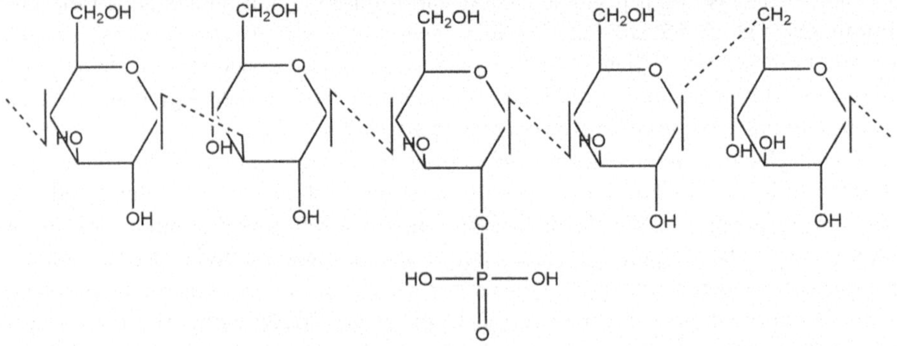


**Fig. 6.5** Structure of phosphorylated β-glucan

water solubility and biological activities since it can self-associate in the form of triple helixes (Williams et al., 1991).

Studies reported that the introduction of a phosphate group in β-glucan significantly enhances the anti-tumor activity. X. Chen et al. (2009) synthesized a water-soluble phosphorylated *poria cocos* β-glucans by using phosphoric acid. The structure, average MW, and chain conformation of phosphorylated β-glucans were studied. It was observed that, at 0.15 M NaOH solution, phosphorylated *poria cocos* β-glucans exhibited higher flexibility and a strong inhibition against S-180 tumor cells. The glucan fraction with high MW exhibited stronger anti-tumor activity even at low dosages. To trigger the anti-tumor activity of polysaccharides, they must collide and bind with the receptor molecules. Thus the polysaccharide with high MW has a higher possibility of collision with the substrate molecule. Similarly, Ye et al. (2013) reported that as compared to native *Lachnum* Sp β-glucan, phosphorylated β-glucan exhibited potential antiproliferative activity in murine sarcoma S-180 cells.

Mei et al. (2020) phosphorylated the yeast cell walls β-glucan using phosphorous oxychloride in the presence of DMSO and anhydrous pyridine. It was noticed that the phosphorylated β-glucan exhibited potential antioxidant activity (DPPH scavenging). They also observed a significant dose-effect relationship between glucan concentration and its ability to scavenge DPPH radicals. High and medium doses of phosphorylated β-glucan increased superoxide dismutase and catalase contents in mice's serum, liver, and brain and reduced malondialdehyde levels in the liver and brain sufficiently.

### 6.3.4 Acetylation

In the acetylation process, the hydroxyl group of saccharides is replaced with the acetyl group (Fig. 6.6). A wide range of commercial applications has been proposed for this process. However, few studies examined how the acetylation of β-glucans affects biological activity. The number of studies on the acetylation of β-glucans is limited, whereas it is widely explored in the modification of other polymers like starch (Francini Yumi Kagimura et al., 2015b). Acetylation is generally performed by adding acetic anhydride under alkaline conditions (pH 8–8.4). According to FDA, the DS of acetylated compounds for food application lies between 0.01 and 0.20.

The acetylation of oat β-glucan with a DS 0.03 and 0.12 significantly increased the glucose availability and decreased the fat binding capacity. In contrast, it enhanced the swelling and bile acid binding capacity. In addition, the gel formed from acetylated β-glucans decreased hardness and did not show any cohesiveness compared to the parental molecule. The rheological studies showed that the acetylated β-glucans have decreased viscosity and showed a viscous behavior (De Souza et al., 2015).

Similarly, a DS-dependent change in the morphology of the acetylated ascomycetes fungus β-glucan structure was noticed by Luna et al. (2018). It changed the

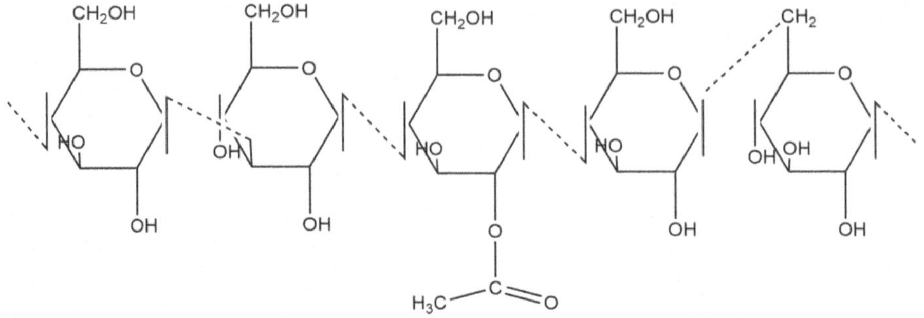


**Fig. 6.6** Structure of acetylated β-glucan

thickness and translucency of β-glucan. Further, it enhanced the antioxidant activity, and it depends on the DS. An acetylated β-glucan with a low DS (0.44) shows higher OH* and $H_2O_2$ scavenging activities and higher reducing power than the parental molecule and acetylated derivatives with a higher DS.

### 6.3.5 *Oxidation*

Oxidation modifies the physicochemical and functional characteristics of polymers and is generally performed by treating the polymer with an oxidizing agent like hydrogen peroxide, sodium hypochlorite, etc. During oxidation treatment, the hydroxyl group of the molecule is oxidized to carboxyl or carbonyl groups. It may lead to the depolymerization of the molecule and alters its functional properties. The oxidation of β-glucans significantly enhances its solubility.

The oat β-glucans oxidation using hydrogen peroxide significantly improved its cholesterol reduction ability (de Moura et al., 2011). In addition, it enhances the cholic acid binding ability and altered the swelling capacity. However, it did not alter the fat-binding ability. Similarly, the rheological and textural characteristics of β-glucans gel were altered by oxidation. Oxidation with $H_2O_2$ significantly reduced gel viscosity, hardness, adhesiveness, and gumminess. Qin et al. (2021) reported that the bile acid binding capacity, antioxidant, and antimicrobial activity of oat β-glucans potentially increased by oxidation using $H_2O_2$.

The oxidation of barley β-glucans with 0.6% $H_2O_2$ resulted in an increase in its solubility from 4.81% to 5.45% (Lee et al., 2016). Oxidation significantly enhanced the purity (35–70%) and oil-binding ability of β-glucans, whereas it decreased the MW, viscosity, and water-binding abilities. Mäkelä et al. (2015) observed the impact of different oxidizing agents, i.e., $H_2O_2$ and ascorbic acid, on the degradation of barley β-glucans. The researchers reported that $H_2O_2$ oxidation is faster than ascorbic acid oxidation. Both $H_2O_2$ and ascorbic acid resulted in the cleavage of glycosidic linkage or cross rings. The oxidation significantly decreased the MW and viscosity, whereas it shows an increased tendency toward agglomeration.

The rheological study of oxidized yeast β-glucans showed that a concentration higher than 5 mg/mL exhibited a shear thinning behavior of the pseudoplastic non-newtonian fluid (Y. Guo et al., 2023). It decreased the fluidity index of β-glucans, indicating an increase in elasticity and viscosity. Similarly, the oxidation of *Poria cocus* β-glucans resulted in significant depolymerization of the glucan chain, and it resulted in enhanced water solubility, bile acid binding capacity, and antioxidant activity (Y. Wang et al., 2011).

### 6.3.6 Acid Hydrolysis

In a high moisture environment, the use of β-glucans with high MW is challenging since they will agglomerate and form a semisolid. The MW must be decreased to obtain a high concentration of β-glucans without forming aggregates. Several methods are adopted to reduce the MW of β-glucans, and the hydrolysis of β-glucans using enzymes or acids are one of the common method used for MW reduction of β-glucans. Ishimoto et al. (2018) observed an increase in solubility and a reduction in MW of yeast β-glucan with a treatment of 45% sulphuric acid at 20 °C.

Johansson et al. (2006) observed the effect of different acids, i.e., hydrochloric, trifluoroacetic, and sulphuric acid, on the hydrolysis of β-glucan. The results revealed that under mild acid concentration, at a condition similar to the human stomach (37 °C and pH 1), only hydrochloric acid carried out the β-glucan degradation or hydrolysis over a 12 h incubation. Sibakov et al. (2013) stated that the acid hydrolysis (hydrochloric acid) of oat bran β-glucan resulted in the breakdown of 1–3 and 1–4 β-D- linkages, and the MW was decreased from 86,000 to 34,000 g/mol. They have also added that the β-glucan degradation can be controlled by controlling the temperature.

The degradation of oat β-glucan with hydrochloric acid significantly altered the physicochemical properties (Qin et al., 2021). The acid-treated β-glucan did not show any swelling capacity but enhanced the fat-binding ability and hypoglycemic activity. In addition, the acid-degraded oat β-glucan showed an enhanced antioxidant and antimicrobial activity against *Staphylococcus aureus*.

## 6.4 Biological Modification

In the biological modification of β-glucan, enzymes are mainly used to catalyze the degradation of β-glucan. The biological methods of modification are eco-friendly and mild. Compared to chemical modification, enzymatic modification is highly specific, efficient, and has fewer side effects. In an enzymatic modification, the rate and direction of the reaction can be controlled by controlling the treatment conditions. The biological modification of β-glucan is generally carried out either by using enzymes or by using microbes.

### 6.4.1 Enzymatic Modification

A wide range of industrial processes uses enzyme modification to change the functionalities of β-glucan. The enzymatic modification mainly results in the depolymerization of β-glucan, and it depends on the specificity of the enzyme used in particular industrial applications. It is sometimes possible to produce specific oligosaccharide fragments from functional polysaccharides by maintaining or slightly altering their molecular size. As a result of enzymatic degradation, polysaccharides are primarily degraded at the backbone, thereby reducing their molecular weight (MW) and viscosity (Yuan et al., 2020). The enzyme-assisted breakdown of β-glucan results in the exposure of a large number of β-1-3 linkages, and reduced particle size lead to enhanced immune activity. Similarly, a decrease in MW of β-glucan was noticed in endogenous β-glucanase treated barley β-glucan (Mio et al., 2020) and wheat β-glucanase treated barley β-glucan (De Arcangelis et al., 2019).

Sibakov et al. (2013) studied the impact of enzymatic hydrolysis on MW of oat β-glucan at a 50% water content. It enabled controlled depolymerization of the β-glucan by the hydrolysis of (1–4) β-D- glycosidic linkages. The enzymatic hydrolysis mainly depends on the treatment time. The treatment of oat β-glucan for 3–4 h reduced the MW from 71,000 to 49,000 g/mol. The β-glucan hydrolysate prepared from oat bran exhibited a hypocholesterolemic effect and *in-vitro* bile acid binding capacity (I. Y. Bae et al., 2010).

### 6.4.2 Microbial Modification

The microbial modification includes changing the structure and composition of a polysaccharide by fermentation using microorganisms, thus altering its physical properties and biological activity (Yuan et al., 2020). Liang et al. (2018) synthesized a water-soluble β-1,3-glucan from curdlan by fermentation using a mixed *Agrobacterium* sp. and *Trichoderma harzianum* culture. The *in-vitro* analysis of β-1,3-glucan shows that at a higher concentration, it exhibits anti-tumor activity against MCF-7, HepG2, and Hela cancer cells.

The fermentation of barley β-glucan by *Lactobacillus plantarum dy-1* resulted in changes in its structure, altering its physiological activities. The fermented β-glucan decreased the α-amylase, α-glycosidase, and lipase activity and decreased cholesterol binding capacity (Xiao et al., 2020).

## 6.5 Conclusion

The β-glucan is one of the most important polysaccharides for human health. Several potential health benefits associated with it have been researched, including chronic diseases, anti-tumor properties, etc. However, depending on the molecular weight (MW) and structure, β-glucans are partially soluble or insoluble in water and other solvents, restricting their wide utility. Different derivatization and MW degradation methods are adopted to enhance its solubility. The derivatization of β-glucan can be achieved by modifying the β-glucan using different chemical methods, like carboxy methylation, sulfation, phosphorylation, oxidation, etc., The MW degradation of β-glucan is mainly achieved by physical modification like thermal degradation, irradiation, etc., The modification of β-glucan by different methods significantly enhances its solubility, and it also enhances its bioactivities.

## References

Adachi, Y., Ohno, N., Ohsawa, M., Oikawa, S., & Yadomae, T. (1990). Change of biological activities of (1→ 3)-β-D-glucan from Grifola frondosa upon molecular weight reduction by heat treatment. *Chemical Pharmaceutical Bulletin, 38*(2), 477–481. https://doi.org/10.1248/cpb.38.477

Babincová, M., Bačová, Z., Machová, E., & Kogan, G. (2002). Anti-oxidant properties of carboxymethyl glucan: Comparative analysis. *Journal of Medicinal Food, 5*(2), 79–83. https://doi.org/10.1089/109662002760178159

Bae, I. Y., Kim, S. M., Lee, S., & Lee, H. G. (2010). Effect of enzymatic hydrolysis on cholesterol-lowering activity of oat β-glucan. *New Biotechnology, 27*(1), 85–88. https://doi.org/10.1016/j.nbt.2009.11.003

Bae, I. Y., Chang, Y. J., Kim, H. W., & Lee, H. G. (2008). Anticoagulant activity of sulfated barley β -glucan. *Food Science and Biotechnology, 17*, 870–872.

Baranoski, A., Tempesta Oliveira, M., Semprebon, S. C., Niwa, A. M., Ribeiro, L. R., & Mantovani, M. S. (2015). Effects of sulfated and non-sulfated β -glucan extracted from Agaricus brasiliensis in breast adenocarcinoma cells -MCF-7. *Toxicology Mechanisms and Methods, 25*(9), 672–679. https://doi.org/10.3109/15376516.2015.1043762

Benchamas, G., Huang, S., & Huang, G. (2020). The influence of traditional and new processing technologies on the structure and function of food polysaccharide. *Food and Function, 11*(7), 5718–5725. https://doi.org/10.1039/d0fo00854k

Byun, E.-H., Kim, J.-H., Sung, N.-Y., Choi, J., Lim, S.-T., Kim, K.-H., Yook, H.-S., Byun, M.-W., & Lee, J.-W. (2008). Effects of gamma irradiation on the physical and structural properties of b-glucan. *Radiation Physics and Chemistry, 77*(6), 781–786. https://doi.org/10.1016/j.radphyschem.2007.12.008

Chen, C., Wu, W., Xu, X., Zhang, L., Liu, Y., & Wang, K. (2014). Chain conformation and antitumor activity of derivatives of polysaccharide from Rhizoma Panacis Japonici. *Carbohydrate Polymers, 105*(1), 308–316. https://doi.org/10.1016/j.carbpol.2014.01.089

Chen, J., Chen, L., Lin, S., Liu, C., & Cheung, P. C. K. (2015). Preparation and structural characterization of a partially depolymerized beta-glucan obtained from Poria cocos sclerotium by ultrasonic treatment. *Food Hydrocolloids, 46*, 1–9. https://doi.org/10.1016/j.foodhyd.2014.12.005

Chen, X., Xu, X., Zhang, L., & Zeng, F. (2009). Chain conformation and anti-tumor activities of phosphorylated (1→3)-β-d-glucan from Poria cocos. *Carbohydrate Polymers, 78*(3), 581–587. https://doi.org/10.1016/j.carbpol.2009.05.019

Chen, X., Zhang, L., & Cheung, P. C. K. (2010). Immunopotentiation and anti-tumor activity of carboxymethylated-sulfated β-(1→3)-d-glucan from Poria cocos. *International Immunopharamacology, 10*, 398–405. https://doi.org/10.1016/j.intimp.2010.01.002

De Arcangelis, E., Djurle, S., Andersson, A. A. M., Marconi, E., Messia, M. C., & Andersson, R. (2019). Structure analysis of β-glucan in barley and effects of wheat β-glucanase. *Journal of Cereal Science, 85*, 175–181. https://doi.org/10.1016/j.jcs.2018.12.002

De Souza, N. L., Bartz, J., Zavareze, E. D. R., De Oliveira, P. D., Da Silva, W. S. V., Alves, G. H., & Dias, A. R. G. (2015). Functional, thermal and rheological properties of oat β-glucan modified by acetylation. *Food Chemistry, 178*, 243–250. https://doi.org/10.1016/j.foodchem.2015.01.079

Faure, A. M., Knüsel, R., & Nyström, L. (2013). Effect of the temperature on the degradation of β-glucan promoted by iron(II). *Bioactive Carbohydrates and Dietary Fibre, 2*(2), 99–107. https://doi.org/10.1016/j.bcdf.2013.09.003

Fu, W., Zhao, G., & Liu, J. (2022). Effect of preparation methods on physiochemical and functional properties of yeast β-glucan. *LWT - Food Science and Technology, 160*, 113284. https://doi.org/10.1016/j.lwt.2022.113284

Fujiike, A. Y., Lee, C. Y. A. L., Rodrigues, F. S. T., Oliveira, L. C. B., Barbosa-Dekker, A. M., Dekker, R. F. H., Cólus, I. M. S., & Serpeloni, J. M. (2022). Anticancer effects of carboxymethylated (1→3)(1→6)-β-D-glucan (botryosphaeran) on multicellular tumor spheroids of MCF-7 cells as a model. *Journal of Toxicology and Environmental Health, Part A, 85*(13), 521–537. https://doi.org/10.1080/15287394.2022.2048153

Ge, Y., Ahmed, S., Yao, W., You, L., Zheng, J., & Hileuskaya, K. (2021). Regulation effects of indigestible dietary polysaccharides on intestinal microflora: An overview. *Journal of Food Biochemistry, 45*(1), 1–10. https://doi.org/10.1111/jfbc.13564

Guo, H., Feng, K.-L., Zhou, J., Liu, L., Wei, S.-Y., Zhao, L., Qin, W., Gan, R.-Y., & Wu, D.-T. (2020). Carboxymethylation of Qingke β-glucans and their physicochemical properties and biological activities. *International Journal of Biological Macromolecules, 147*, 200–208. https://doi.org/10.1016/j.ijbiomac.2020.01.050

Guo, H., Li, H. Y., Liu, L., Wu, C. Y., Liu, H., Zhao, L., Zhang, Q., Liu, Y. T., Li, S. Q., Qin, W., & Wu, D. T. (2019). Effects of sulfated modification on the physicochemical properties and biological activities of β-glucans from Qingke (Tibetan hulless barley). *International Journal of Biological Macromolecules, 141*, 41–50. https://doi.org/10.1016/j.ijbiomac.2019.08.245

Guo, Y., Ma, H., Cai, W., & Huang, Q. (2023). Oxidized yeast β-glucan: Rheological behaviors and the formation of entanglement network at different oxidation degree. *Food Hydrocolloids, 137*, 108364. https://doi.org/10.1016/j.foodhyd.2022.108363

Hu, T., Cai, W., Cai, W., Zheng, Z., Xiao, Y., & Huang, Q. (2021). Structure, size and aggregated morphology of a β-D-glucan from Lignosus rhinocerotis as affected by ultrasound. *Carbohydrate Polymers, 269*. https://doi.org/10.1016/j.carbpol.2021.118344

Ishimoto, Y., Ishibashi, K. I., Yamanaka, D., Adachi, Y., Kanzaki, K., Iwakura, Y., & Ohno, N. (2018). Production of low-molecular weight soluble yeast β-glucan by an acid degradation method. *International Journal of Biological Macromolecules, 107*, 2269–2278. https://doi.org/10.1016/j.ijbiomac.2017.10.094

Ishimoto, Y., Ishibashi, K. I., Yamanaka, D., Adachi, Y., Kanzaki, K., Okita, K., Iwakura, Y., & Ohno, N. (2017). Modulation of an innate immune response by soluble yeast β-glucan prepared by a heat degradation method. *International Journal of Biological Macromolecules, 104*, 367–376. https://doi.org/10.1016/j.ijbiomac.2017.06.036

Johansson, L., Virkki, L., Anttila, H., Esselström, H., Tuomainen, P., & Sontag-Strohm. (2006). Hydrolysis of β-glucan. *Food Chemistry, 97*, 71–79. https://doi.org/10.1016/j.foodchem.2005.03.031

Kagimura, F. Y., da Cunha, M. A. A., Theis, T. V., Malfatti, C. R. M., Dekker, R. F. H., Barbosa, A. M., Teixeira, S. D., & Salomé, K. (2015a). Carboxymethylation of (1→6)-β-glucan (lasiodiplodan)_ preparation, characterization and anti-oxidant evaluation. *Carbohydrate Polymers, 127*, 390–399. https://doi.org/10.1016/j.carbpol.2015.03.045

Kagimura, F. Y., da Cunha, M. A. A., Barbosa, A. M., Dekker, R. F. H., & Malfatti, C. R. M. (2015b). Biological activities of derivatized d-glucans: A review. *International Journal of Bio, 72*, 588–598. https://doi.org/10.1016/j.ijbiomac.2014.09.008

Kaur, R., Sharma, M., Ji, D., Xu, M., & Agyei, D. (2020). Structural features, modification, and functionalities of beta-glucan. *Fibers, 8*(1), 1–29. https://doi.org/10.3390/fib8010001

Khan, A. A., Gani, A., Masoodi, F. A., Amin, F., Wani, I. A., Khanday, F. A., & Gani, A. (2016). Structural, thermal, functional, anti-oxidant & antimicrobial properties of β-d-glucan extracted from baker's yeast (Saccharomyces cereviseae)—Effect of γ-irradiation _ Elsevier Enhanced Reader.pdf. *Carbohydrate Polymers, 140*, 442–450. https://doi.org/10.1016/j.carbpol.2016.01.003

Lee, S. H., Jang, G. Y., Kim, M. Y., Hwang, I. G., Kim, H. Y., Woo, K. S., Lee, M. J., Kim, T. J., Lee, J., & Jeong, H. S. (2016). Physicochemical and in vitro binding properties of barley β-glucan treated with hydrogen peroxide. *Food Chemistry, 192*, 729–735. https://doi.org/10.1016/j.foodchem.2015.07.063

Liang, Y., Zhu, L., Gao, M., Wu, J., & Zhan, X. (2018). Effective production of biologically active water-soluble β-1,3-glucan by a coupled system of Agrobacterium sp. and Trichoderma harzianum. *Preparative Biochemistry and Biotechnology, 48*(5), 446–456. https://doi.org/10.1080/10826068.2018.1452259

Liu, H., Li, Y., Gao, J., Shi, A., Liu, L., Hu, H., Putri, N., Yu, H., Fan, W., & Wang, Q. (2016). Effects of microfluidization with ionic liquids on the solubilization and structure of β-D-glucan. *International Journal of Biological Macromolecules, 84*, 394–401. https://doi.org/10.1016/j.ijbiomac.2015.12.014

Liu, Y., Liu, Y., Jiang, H., Xu, L., Cheng, Y., Wang, P. G., & Wang, F. (2014). Preparation, antiangiogenic and antitumoral activities of the chemically sulfated glucan from Phellinus ribis. *Carbohydrate Polymers, 106*(1), 42–48. https://doi.org/10.1016/j.carbpol.2014.01.088

Luna, W. N. S., Santos, V. A. Q., Teixeira, S. D., Barbosa-Dekker, A. M., Dekker, R. F. H., & da Cunha, M. A. A. (2018). O-acetylated (1→6)-β-D-glucan (Lasiodiplodan): Chemical derivatization, characterization and anti-oxidant activity. *Journal of Pharmacy and Pharmacology, 6*, 320–332. https://doi.org/10.17265/2328-2150/2018.04.003

Mäkelä, N., Sontag-Strohm, T., & Maina, N. H. (2015). The oxidative degradation of barley β-glucan in the presence of ascorbic acid or hydrogen peroxide. *Carbohydrate Polymers, 123*, 390–395. https://doi.org/10.1016/j.carbpol.2015.01.037

Mei, X., Tang, Q., Huang, G., Long, R., & Huang, H. (2020). Preparation, structural analysis and anti-oxidant activities of phosphorylated (1 → 3)-β-D-glucan. *Food Chemistry, 309*, 125791. https://doi.org/10.1016/j.foodchem.2019.125791

Mio, K., Yamanaka, C., Ichinose, Y., Kohyama, N., Yanagisawa, T., & Aoe, S. (2020). Effects of barley β-glucan with various molecular weights partially hydrolyzed by endogenous β-glucanase on glucose tolerance and lipid metabolism in mice. *Cereal Chemistry, 97*(5), 1056–1065. https://doi.org/10.1002/cche.10328

de Moura, F. A., Pereira, J. M., da Silva, D. O., Zavareze, E. d. R., Moreira, A. d. S., Helbig, E., & Dias, A. R. G. (2011). Effects of oxidative treatment on the physicochemical, rheological and functional properties of oat β-glucan. *Food Chemistry, 1128*, 982–987. https://doi.org/10.1016/j.foodchem.2011.04.003

Ohno, N., Kurachi, K., & Yadomae, T. (1988). Physicochemical properties and anti-tumor activities of carboxymethylated dericatives of glucan from Sclerotinia sclerotiorum. *Chemical Pharmaceutical Bulletin, 36*, 1016–1025. https://doi.org/10.1248/cpb.36.1016

Oliveira, L. d. C., Gutkoski, L. C., Elias, M. C., Mazzutti, S., Aosani, E., & da Rocha, J. C. (2010). Effect of drying temperature on the quality of white oat grains. *Ciencia e Agrotecnologia, 34*(2), 313–319. https://doi.org/10.1590/s1413-70542010000200007

Qin, Y., Xie, J., Xue, B., Li, X., Gan, J., Zhu, T., & Sun, T. (2021). Effect of acid and oxidative degradation on the structural, rheological, and physiological properties of oat β-glucan. *Food Hydrocolloids, 112*, 106284. https://doi.org/10.1016/j.foodhyd.2020.106284

Shah, A., Ahmad, M., Ashwar, B. A., Gani, A., Masoodi, F. A., Wani, I. A., Wani, S. M., & Gani, A. (2015a). Effect of γ-irradiation on structure and nutraceutical potential of β-d-glucan from barley (Hordeum vulgare). *International Journal of Biological Macromolecules, 72*, 1168–1175. https://doi.org/10.1016/j.ijbiomac.2014.08.056

Shah, A., Masoodi, F. A., Gani, A., & Ashwar, B. A. (2015b). Effect of γ-irradiation on antioxidant and antiproliferative properties of oat β-glucan. *Radiation Physics and Chemistry, 117*, 120–127. https://doi.org/10.1016/j.radphyschem.2015.06.022

Shun, C. W., Bligh, S. W. A., Chang, L. Z., Song, S. S., Zheng, T. W., Zhi, B. H., Crowder, J., Branford-White, C., & Vella, C. (2008). Sulfated β-glucan derived from oat bran with potent anti-HIV activity. *Journal of Agricultural and Food Chemistry, 56*(8), 2624–2629. https://doi.org/10.1021/jf072888h

Sibakov, J., Myllymäki, O., Suortti, T., Kaukovirta-Norja, A., Lehtinen, P., & Poutanen, K. (2013). Comparison of acid and enzymatic hydrolyses of oat bran β-glucan at low water content. *Food Research International, 52*(1), 99–108. https://doi.org/10.1016/j.foodres.2013.02.037

Song, J., Chen, H., Wei, Y., & Liu, J. (2020). Synthesis of carboxymethylated β-glucan from naked barley bran and its antibacterial activity and mechanism against Staphylococcus aureus. *Carbohydrate Polymers, 242*, 116418. https://doi.org/10.1016/j.carbpol.2020.116418

Sousa, F. T. G. d., Biering, S. B., Patel, T. S., Blanc, S. F., Camelini, C. M., Venzke, D., Nunes, R. J., Romano, C. M., Beatty, P. R., Sabino, E. C., & Harris, E. (2022). Sulfated β-glucan from Agaricus subrufescens inhibits flavivirus infection and nonstructural protein 1-mediated pathogenesis. *Anti-viral Research, 203*, 105330. https://doi.org/10.1016/j.antiviral.2022.105330

Sung, N. Y., Byun, E. H., Kwon, S. K., Song, B. S., Choi, J. I., Kim, J. H., Byun, M. W., Yoo, Y. C., Kim, M. R., & Lee, J. W. (2009). Immune-enhancing activities of low molecular weight β-glucan depolymerized by gamma irradiation. *Radiation Physics and Chemistry, 78*(7–8), 433–436. https://doi.org/10.1016/j.radphyschem.2009.03.022

Tao, Y., & Xu, W. (2008). Microwave-assisted solubilization and solution properties of hyper-branched polysaccharide. In *Carbohydrate research* (Vol. 343(18), pp. 3071–3078). https://doi.org/10.1016/j.carres.2008.09.009

Taubner, T., Marounek, M., & Synytsya, A. (2020). Preparation and characterization of hydrophobic and hydrophilic amidated derivatives of carboxymethyl chitosan and carboxymethyl β-glucan. *International Journal of Biological Macromolecules, 163*, 1433–1443. https://doi.org/10.1016/j.ijbiomac.2020.07.257

Wang, M., Zhang, L., Yang, R., Fei, C., Wang, X., Zhang, K., Wang, C., Zheng, W., & Xue, F. (2016). Improvement of immune responses to influenza vaccine (H5N1) by sulfated yeast beta-glucan. *International Journal of Biological Macromolecules, 93*, 203–207. https://doi.org/10.1016/j.ijbiomac.2016.06.057

Wang, Q., Sheng, X., Shi, A., Hu, H., Yang, Y., Liu, L., Fei, L., & Liu, H. (2017). β-glucans: Relationships between modification, conformation and functional activities. *Molecules, 22*(2), 257. https://doi.org/10.3390/molecules22020257

Wang, Y., & Zhang, L. (2006). Chain conformation of carboxymethylated derivatives of (1 → 3)-β-d-glucan from Poria cocos sclerotium. *Carbohydrate Polymers, 65*, 504–509. https://doi.org/10.1016/j.carbpol.2006.02.014

Wang, Y., Liu, S., Yang, Z., Zhu, Y., Wu, Y., Huang, J., & Mao, J. (2011). Oxidation of β-glucan extracted from Poria Cocos and its physiological activities. *Carbohydrate Polymers, 85*(4), 798–802. https://doi.org/10.1016/j.carbpol.2011.03.052

Wang, Y., Yu, Y., & Mao, J. (2009). Carboxymethylated β-glucan derived from Poria cocos with biological activities. *Journal of Agricultural and Food Chemistry, 57*, 10913–10915. https://doi.org/10.1021/jf902589m

Williams, D. L., Pretus, H. A., McNamee, R. B., Jones, E. L., Ensley, H. E., William Browder, I., & Di Luzio, N. R. (1991). Development, physicochemical characterization and preclinical efficacy evaluation of a water soluble glucan sulfate derived from Saccharomyces cerevisiae. *Immunopharmacology, 22*(3), 139–156. https://doi.org/10.1016/0162-3109(91)90039-2

Wu, L., Zhao, J., Zhang, X., Liu, S., & Zhao, C. (2021). Anti-tumor effect of soluble β-glucan as an immune stimulant. *International Journal of Biological Macromolecules, 179*, 116–124. https://doi.org/10.1016/j.ijbiomac.2021.02.207

Xiao, X., Tan, C., Sun, X., Zhao, Y., Zhang, J., Zhu, Y., Bai, J., Dong, Y., & Zhou, X. (2020). Effects of fermentation on structural characteristics and in vitro physiological activities of barley β-glucan. *Carbohydrate Polymers, 231*, 115685. https://doi.org/10.1016/j.carbpol.2019.115685

Xu, S., Xu, X., & Zhang, L. (2013). Effect of heating on chain conformation of branched β-glucan in water. *Journal of Physical Chemistry B, 117*(28), 8370–8377. https://doi.org/10.1021/jp403202u

Ye, M., Yuan, R. Y., He, Y. L., Du, Z. Z., & Ma, X. J. (2013). Phosphorylation and anti-tumor activity of exopolysaccharide from Lachnum YM120. *Carbohydrate Polymers, 97*(2), 690–694. https://doi.org/10.1016/j.carbpol.2013.05.033

Yuan, H., He, Y., Zhang, H., & Ma, X. (2022). Ultrasound-assisted enzymatic hydrolysis of yeast β-glucan catalyzed by β-glucanase: Chemical and microstructural analysis. *Ultrasonics Sonochemistry, 86*. https://doi.org/10.1016/j.ultsonch.2022.106012

Yuan, H., Lan, P., He, Y., Li, C., & Ma, X. (2020). Effect of the modifications on the physicochemical and biological properties of β-glucan—A critical review. *Molecules, 25*(57), 1–21. https://doi.org/10.3390/molecules25010057

Yucel Falco, C., Sotres, J., Rascón, A., Risbo, J., & Cárdenas, M. (2017). Design of a potentially prebiotic and responsive encapsulation material for probiotic bacteria based on chitosan and sulfated β-glucan. *Journal of Colloid and Interface Science, 487*, 97–106. https://doi.org/10.1016/j.jcis.2016.10.019

Zechner-Krpan, V., Petravić-Tominac, V., Galović, P., Galović, V., Filipović-Grčić, J., & Srečec, S. (2010). Application of different drying methods on β-glucan isolated from spent brewer's yeast using alkaline procedure. *Agriculturae Conspectus Scientificus, 75*(1), 45–50.

Zhang, M., Cheung, P. C. K., Ooi, V. E. C., & Zhang, L. (2004a). Evaluation of sulfated fungal β-glucans from the sclerotium of Pleurotus tuber-regium as a potential water-soluble anti-viral agent. *Carbohydrate Research, 339*, 2297–2301. https://doi.org/10.1016/j.carres.2004.07.003

Zhang, M., Cheung, P. C. K., Zhang, L., Chiu, C.-M., & Ooi, V. E. C. (2004b). Carboxymethylated β-glucans from mushroom sclerotium of Pleurotus tuber-regium as novel water-soluble anti-tumor agent. *Carbohydrate Polymers, 57*, 319–325. https://doi.org/10.1016/j.carbpol.2004.05.008

Zhang, M., Zhang, L., & Cheung, P. C. K. (2003). Molecular mass and chain conformation of carboxymethylated derivatives of β-glucan from sclerotia of Pleurotus tuber-regium.pdf. *Biopolymers, 68*, 150–159. https://doi.org/10.1002/bip.10277

Zhang, Q., Xie, J., Xue, B., Li, X., Gan, J., Bian, X., Qin, Y., & Sun, T. (2021). Effect of sulfated modification on rheological and physiological properties of oat β-glucan oligosaccharides prepared by acid or oxidative degradation. *Journal of Cereal Science, 99*, 103209. https://doi.org/10.1016/j.jcs.2021.103209

Zheng, Z., Huang, Q., Luo, X., Xiao, Y., Cai, W., & Ma, H. (2019). Effects and mechanisms of ultrasound- and alkali-assisted enzymolysis on production of water-soluble yeast B-glucan. *Bioresource Technology, 273*, 394–403. https://doi.org/10.1016/j.biortech.2018.11.035

# Chapter 7
# Beta-Glucan Based Bioactive Delivery Systems

**Jyoti Gohri, Mudasir Bashir Mir, Shubham Rohilla, and Maanas Sharma**

## 7.1 Introduction

β-Glucan is a soluble and non-digestible homopolysaccharide. It is a dietary fibre mostly found in cereals, barley, seaweed and various bacteria as well as fungi such as *Aspergillus fumigatus, Candida albicans* and *Saccharomyces cerevisiae* etc. (Ye et al., 2022). β-glucan accumulates in the inner portion of yeast's cell wall and the aleurone portion of cereals, where it aids in the formation of the cell wall. β-Glucan consists of over 250,000 D-glucopyranose units which are linked through glycosidic bonds (Vannucci et al., 2013). β-(1 → 3)-bond is the backbone of the structure with side chains attached that could be either β-(1 → 4) in the case of cereals or β-(1 → 6) in the case of fungi (Han et al., 2020). There is also another complex structure containing cyclic β –(1 → 3, 1 → 6) and hyperbranched β-(1 → 3, 1 → 4, 1 → 6) present in a bacteria belonging to the *Rhizobiaceae* family and *Pleurotus tuber-region* mushroom sclerotia (Breedveld & Miller, 1994). The structure varies according to the type of molecular structure, solubility, type of linkage, and chain conformation depending on the source. In a study on wild barley and farmed barley (Nishantha et al., 2018) found the amounts of β-glucan to be 3.27–6.67% and

J. Gohri
Department of Food Technology and Nutrition, School of Agriculture, Lovely Professional University, Phagwara, Punjab, India

M. B. Mir
Department of Food Science & Technology, Govt. Degree College, Baramulla, Jammu & Kashmir, India

S. Rohilla
Indian Institute of Packaging – Delhi, Patparganj, Delhi, India

M. Sharma (✉)
Department of Food Technology and Home Science, Vivekananda College, University of Delhi, Vivek Vihar, Delhi, India

 
M. Bashir Mir, S. A. Mir (eds.), *Beta-Glucan: Sources, Properties and Applications*,
https://doi.org/10.1007/978-3-031-95788-8_7

2.68–4.74%, respectively. It was discovered that waxy type barley or barley with the hull present had greater levels of β-glucan than hulled and non-waxy barley (Islam et al., 2021). By merging Glacier AC 38 with the Shikoku hadaka type of barley, the Rural Development Administration (RDA) of Korea has created a new variety of barley known as Betaone, which has a greater concentration of β-glucan than previous kinds.

β-glucan has become popular over time because of its beneficial role in human health, it is an excellent immune stimulant capable of enhancing the innate and immune systems. Combining β-glucan with inactivated CyHV-2 vaccine improved the immune response against gibel carp and other fish diseases (M. Liu et al., 2018b). β-glucan is a great source for people suffering from diabetes as it helps maintain blood cholesterol levels and glucose and triglyceride levels. On ingesting β-glucan, enzyme inactivation occurs along with an increase in viscosity of the small intestine which results in the reduction in absorption of bile acid and cholesterol therefore decreasing plasma cholesterol (Murphy et al., 2020). Oat bran meal treated with 80% ethanol for 30–120 min produced enzyme-deactivated oat bran flour with phenolic compounds which are discarded as waste (Emmons & Peterson, 1999; Wood et al., 1978). Consuming 3 g per day of β-glucan reduces the chances of cardiovascular diseases as recommended by FDA. It also reduces hypertension and prevents diseases such as coronary heart disease, and colon cancer. *Ganaderma lucidum* is abundant in β-glucan, it is used in chemotherapy for treating cancer (Sudhakar et al., 2021). The honeycomb structure of β-glucan allows it to hold bioactive compounds that are resistant to heat, enzymes, and acid. A diet with high levels of b-glucan can help reduce weight (Emmons & Peterson, 1999; Wood et al., 1978). Consuming 3 g per day of β-glucan reduces the chances of cardiovascular diseases as recommended by FDA. It also reduces hypertension and prevents. Due to the property of β-glucan of biosynthesis induction of collagen, the healing process of wounds gets faster leading to the invention of stronger wound dressings, not only this it is also used as anti-infective drug and vaccine adjuvants in human body (Desai et al., 2023). The honeycomb structure of β-glucan allows it to hold bioactive compounds that are resistant to heat, enzymes, and acid. A diet with high levels of β-glucan can help reduce weight in over 8–12 weeks. Due to its stability and safety for human consumption, β-glucan has been designated as General Recognised as Safe (GRAS) by the US Food and Drug Administration (FDA).

Apart from being beneficial for human health, In the realms of commerce, nutrition, and nutraceuticals, β-glucan's physical and functional qualities including thickening, stabilising, emulsification, and gelation are crucial. Chemically it is used in nanoparticle synthesis for drug delivery. β-glucan being a hydrocolloid can manipulate the textural properties along with the rheological properties of the food materials. It is used in the preparation of soups, sauces, and beverages. Adding β-glucan to bread and cake preparation can enhance the physiological properties and enhance the quality and quantity of loaves of bread. In pasta products, β-glucan showed low glycaemic index which made it effective against metabolic diseases (Chillo et al., 2011). Hyperglycaemic condition could be prevented during bread preparation by the inclusion of β-glucan which slows the glucose release leading to enhanced digestion of starch. β-glucan is also used in beverage industries such as producing ice cream and yoghurt containing low fat level (Abdel-Haleem & Awad, 2015). Further, there are limitations to the native β-glucan like high viscosity, low solubility, less bioavailability, low stability at low temperatures and less permeability in cell, these are due to the large size of β-glucan that leads to a slow release of the encapsulated material in its matrix (ul Ashraf et al., 2020). This could be easily overcome by reducing the size of β-glucan to a nano level using fabrication techniques, the precise size improves the solubility, functioning, physiochemical and nutraceutical properties. Strong adhesive forces are created when the size is reduced from micrometres to manometers, which increases the bioavailability of the substance that is encapsulated. Also, the control over the delivery systems could be handled in a better format which would result in higher incorporation into the food materials without affecting other properties such as sensory properties (López-Rubio & Lagaron, 2012).

## 7.2 The Role of β-Glucan in Bioactive Delivery

### *7.2.1 Encapsulation of Seabuckthorn Oil in β-Glucans*

Sea buckthorn oil is regarded for its high concentration of polyunsaturated fatty acids, carotenoids, polyphenols, and tocopherols; nevertheless, its sensitivity to sunlight, oxygen, and microbial activity limits its application in food products such as yogurt, bread, and beverages. Microencapsulation is a promising approach for enclosing oil in protective wall materials, increasing its stability and shelf life. This method avoids lipid deterioration, since studies demonstrate that oil encapsulated with 3% beta-glucan had much reduced lipid oxidation (5.50 mEq/kg fat) after 1 week than oil kept conventionally (31.78 mEq/kg fat). Beta-glucan, a soluble dietary fiber, has shown promise as a coating material, enhancing emulsion stability and encapsulation efficiency. Microcapsules containing 0.5% beta-glucan exhibited reduced particle sizes, 72.2% encapsulation efficiency, low humidity (2.47%), and low density (0.52 g/cm$^3$). However, greater concentrations (1.0%) decreased effectiveness due to bigger particle sizes. Beta-glucan also has health advantages and can be added to foods such as bread, pasta, and yogurt to enhance fiber intake while retaining the characteristics of sea buckthorn oil. Thus, microencapsulation with beta-glucan improves oil stability, broadens its functional food applications, and addresses consumer demand for health-promoting additives (Drozińska et al., 2019).

### *7.2.2 Antioxidants from Tocopherols*

Polyunsaturated fatty acids (PUFAs) and α-tocopherol were encapsulated in powders with maltodextrin (Dextrose Equivalent, DE 12) as the wall material and several emulsifiers—Tween 20, acacia gum, and β-glucan-rich extracts from Pleurotus ostreatus. The study evaluated how emulsifiers affect oil droplet size, α-tocopherol stability during in vitro digestion, and oxidative stability during accelerated storage. Acacia gum-stabilized emulsions had consistent particle size distribution until the stomach phase, but showed considerable α-tocopherol degradation before the intestinal stage. In contrast, β-glucan-stabilized emulsions had a bimodal droplet size distribution in the oral and gastric phases, but preserved α-tocopherol in the intestinal phase. However, α-tocopherol was undetectable in all samples after digestion, indicating oxidative destruction. At 60 °C, β-glucan-stabilized powders retained 82% α-tocopherol after 5 days, corresponding to 310 days at 20 °C. Acacia gum and Tween 20 degraded faster. β-glucans have antioxidant characteristics that prevent oxidation and stabilize oil droplets during digestion. However, more research is needed to understand α-tocopherol breakdown in the intestinal phase and its fate in vivo. These findings show that β-glucans are attractive emulsifying agents for encapsulation systems in food applications (Gallotti et al., 2020b).

### *7.2.3 β-Glucan as a Wall Material for Targeted Delivery*

Rutin (Ru) was nano-encapsulated in β-glucan matrices from dates [En-Ru(D)] and mushrooms [En-Ru(M)] utilizing eco-friendly ultrasonication technique to boost its bioavailability, stability, and biological activity in the gastrointestinal system. The nanoparticles' hydrodynamic diameters were 314.04 nm and 482.21 nm, with polydispersity indices of 0.21 and 0.33, respectively. The high encapsulation effectiveness of 89% for date β-glucan and 91% for mushroom β-glucan indicates significant connections between rutin's hydroxyl groups and the β-glucan matrix. FT-IR analysis indicated that the nanoparticles successfully entrapped rutin. The in vitro release followed the Higuchi paradigm, and both En-Ru(D) and En-Ru(M) demonstrated increased rutin bioavailability in intestinal fluid. Encapsulation shielded rutin from the acidic stomach environment and allowed for controlled release in the gastrointestinal system. Encapsulated rutin maintained its anti-obesity and antioxidant properties ($p < 0.05$) and demonstrated antibacterial activity against *E. coli* (ATCC 25922) and *Staphylococcus aureus*. The β-glucan matrix provided prebiotic effects, enhancing rutin's multifunctionality. These findings suggest that β-glucan-based encapsulation can effectively transport flavonoids, leading to targeted release and increased bioactivity. Future studies should focus on introducing rutin-loaded β-glucan nanoparticles into food systems for greater health benefits (Shah et al., 2022).

## 7.3 Types of β-Glucan-Based Delivery Systems

Various β-glucan have been discovered based on their different functional and compositional properties from different sources such as yeast, mushroom, barley, oats or some bacteria. Some of them with their respective sources are LNT (Lentinan), glomerellan (*Glomerella cingulate*-1,3–1.6-linked glucan), curdlan (*Alcaligenes faecalis*-β-1,3-glucan), betafectin (*Saccharomyces cerevisiae*-β-(1,6) branched β-(1,3) glucan), zymocel (*Saccharomyces cerevisiae)* (Ozcan et al., 2017).

Functional properties are greatly affected by physiochemical characteristics such as branching, solubility, extraction methods etc. Different arrangements could lead to different types of interactions with other molecules, triple helical structure of β-glucan could reversible interconversion between single strands coil resulting in the triple-strand helix structure associating itself with polynucleotides and incorporating other molecules in the structure (Y. Zhang et al., 2013). Different delivery systems have been incorporated with different chemical structures and properties which are explained further: -.

### 7.3.1 *Yeast β-Glucan Based Delivery Systems*

Yeast β-glucan has a linear β-1,3 glycosidic linkage backbone with β-1,6 side branches. Its hollow, porous spherical form allows it to carry therapeutic genes to immune cells while encouraging receptor-mediated endocytosis, which aids in cellular absorption (H. Chen et al., 2022c). β-glucan tends to bind with Dectin-1 and receptors present on the macrophage's surfaces, dendritic cells, and phagocytic cells. Dectin-1 endocytosis internalizes these β-glucan hence making it useful for immunomodulatory activity regulated by the pathogen associated molecular pattern (PAMP) (D. H. Lee & Kim, 2014).

Yeast cell microcapsule serves as a novel transporter for small molecules and drugs with low solubility. Positively charged particles were inserted by electrostatic force from the exterior surface easily and negatively charged were incorporated in combination with polycations (Filipczak et al., 2020). A delivery system with an inner shell composed of polymer-lipid hybrid nanoparticles (NPs) and a hollow, porous outer shell built of yeast-based β-glucan microparticles has been developed (Table 7.1). It focused on intestinal phagocytic cells and improved the availability of cabazitaxel for oral transfer (Su et al., 2021). In research a novel amorphous solid dispersion was created, it was based on the particles of yeast glucan with a size of 2–4 μm. Water-soluble drugs such as curcumin and ibuprofen were spray dried in the glucan particle imparting a hydrophilic nature (Milea et al., 2020). This could result in the yeast cell microcapsule internalising quickly in the gut-associated lymphoid tissue (GALT) as shown in Table 7.2. This speeds up the lymphatic system's delivery of macrophages to the locations of inflammation and tumour cells (Houacine et al., 2018; Tan et al., 2022).

### 7.3.2 *Non-yeast β-Glucan Based Delivery Systems*

Non-yeast β-glucan has β-1,3 as the main chain with no side chains that take part in the complex formation. Schizophyllan glucan (SPG) consists of β-1,3 main chain with β-1,6 stem at every three glucose residues (Sumiya et al., 2020). When it reacts

**Table 7.1** β-glucan delivery systems

| Source | Drugs/agent | Cell/model | result | References |
|---|---|---|---|---|
| Yeast derived microcapsule | Positively charged nanotherapies and probes | Mice lacking apolipoprotein E that already have atherosclerotic lesions | Oral delivery of a combination of anti-atherosclerotic nanotherapies packed in YCs resulted in markedly potentiated efficacies in E-deficient animals. | (Su, 2021) |
| β-D-glucan micro-particles | Doxorubicin | MDA-MB-231 cells & HUVEC cells | Exhibiting delayed and sustained release using chitosan and alginate outer layer, and providing synergistic effects for the treatment of tumours with a little dose of DOX. | (Su, 2021) |
| Baker's yeast glucan particles | Ibuprofen & curcumin | – | Poorly water-soluble medications now have better stability and loading capacity thanks to spray drying. | (Milea et al., 2020) |
| Polysaccharides from Pleurotus tuber regium, Polyporus rhinocerus, Coriolus versicolor & Ganoderma lucidum | Gold nanorods | Cell lines MCF-7, SW480, and HT-29. Athymic female Balb/c mice | Low cytotoxicity and great photothermal stability make PTR sclerotial -D-glucan attractive photothermal agents. | (Kumar et al., 2022) |

with poly dA of the therapeutic oligonucleotide (ODN), the CpG moiety gets inserted in the delivery system without much modification or any type of chemical reaction of the polysaccharide (Mochizuki & Sakurai, 2009). SPG/dA-ODN is a better carrier as compared to alone ODN as it gets carried to endosomal compartments with enhanced cellular uptake by the Dectin-1 pathways of the murine peritoneal macrophages (Su et al., 2021). SPG/TNF-α antisense ODN reduces inflammation caused by LPS-induced hepatitis by getting transported into the peritoneal macrophages through Dectin-1, it also induces therapeutic effect when taken efficiently in macrophages with dextran sodium sulphate induced colitis (Su et al., 2021).

Lentinan (LNT) had two β-(1,6) chain at every fifth linear β-(1,3)-linkage. It has high water solubility and was used as a delivery system for inorganic nanoparticles along with gene medicine (Q. Liu et al., 2017). Since the CpG/LNT delivery method has a cleavable disulphide bond, it is effective for gene transfection and has a sensitive release of therapeutic ODN in the presence of reducing agents, such as tumour tissue (Teo et al., 2016).

**Table 7.2** Bioactive encapsulation in glucan based & allied systems

| Sources | Compound | Cell/application | Results | References |
|---|---|---|---|---|
| Probiotic microcapsules | Inulin | Probiotic carrier | Lactobacillus acidophilus microencapsulated with inulin had good encapsulation efficacy (96.75%) and improved vitality 120 days when kept under varied conditions. | (Bi et al., 2022) |
| Zymosan nanoparticles | ssDNA | – | When ssDNA is inserted into the triple helix structure of glucan, DNA-incorporated - D-glucan nanoparticles are created. | (Su, 2021) |
| Yeast cell wall micro-particles | Cabazitaxel | SD rats and uncultured 264.7 cells are used to assess the intestine's absorption location. | a technique for delivering drugs via nano-in-micro that is intended to target intestine phagocytic cells | (Su, 2021) |
| Probiotic microcapsules | IMOS, XOS, FOS and GOS | Probiotic carrier | Storage stability: High viability after 30 days at 4 °C.<br>Simulated digestion: Improved survival in stomach and intestinal fluids.<br>Encapsulation efficiency: Yield of 79.52–89.75%.<br>Co-encapsulation: Enhanced by oligosaccharides. FOS gave the best performance. | (Neri-Numa et al., 2020) |
| Microcapsules | Conjugated Xylo-oligosaccharides with whey protein isolate | Lycopene | Enhanced thermal/storage stability and solubility of the bioactive | (Falsafi et al., 2022) |

(continued)

**Table 7.2** (continued)

| Sources | Compound | Cell/application | Results | References |
|---|---|---|---|---|
| β-D-glucan micro-particles | Doxorubicin | MDA-MB-231 cells & HUVEC cells | Core-Shell design: Chitosan core with alginate outer layer.<br>Drug release: Delayed and sustained release profile.<br>Synergistic action: Enhanced cancer therapy efficacy.<br>Low DOX usage: Effective with minimal drug dosage.<br>Benefits: Biocompatible, reduced side effects, targeted delivery. | (Su et al., 2021) |
| Electrostatic nanocomplexes | Carboxymethyl inulin | Bovine serum albumin | Globulous soluble particles (150 nm) and amorphous insoluble complexes are produced at pH values of 2.60 and 4.00, respectively. | (Qureshi et al., 2023) |
| Microparticles | Inulin | Oregano extract | 66% encapsulation efficiency | (Fraj et al., 2019) |
| Electrostatic nano complexes | Inulin | Resveratrol and probiotics | Globular nanocomplexes (10–100 m) are made utilising an internal gelation process. | (Pateiro et al., 2021) |
| Spray-dried microparticles | Inulin | Oleuropein | Improved availability of bioactive. High bio-efficacy | (Xiao et al., 2022) |

Curdlan is a linear chain structure with a molecular weight of more than $1 \times 10^6$ Da, it precipitates out during the renaturing process. Due to Curdlan's propensity for binding to Dectin-1, it improved tissue transport and receptor-mediated endocytosis. It also has a gelling property which enables it to form a self-assembled hydrogel nanoparticle with have inner part as hydrophobic part and outer covering as hydrophilic segment (Hennink & van Nostrum, 2012). It has weak water solubility which was rectified by modifying the C6 hydroxyl group by adding the sulphate group although these made the structure unstable due to electrostatic repulsion hence salts were used as an alternative solution (Ng et al., 2015). Several chemical modifications were done to enhance the water solubility and therapeutic properties. In amphiphilic curdlan, water soluble carboxylic part was modified with deoxycholic acid forming a spherical nanoparticle with dose and pH dependent drug transfer, it is used as nano-delivery for doxorubicin antitumor drug (Akagi et al., 2012; Su, 2021).

### 7.3.3 Hyper-Branched β-Glucan Based Delivery Systems

Hyperbranched β-glucan originates from various sources such as mushrooms, it is considered a strong nano-carriers due to excellent biocompatibility and loading capacity. *Polyporus rhinocerus* created 70% of β-glucan in dry weight, Cold alkali extract using micronized sonication included glucan with 1,3;1,6 linkage and 0.85 branching (C. Liu et al., 2018a). *Ganoderma sinense* gave a water soluble hyper-branched β-glucan with a β-1,6-linkage and β-1,3, β-1,4 glucopyranosyl side chains (Wang et al., 2016). These different structures enable hyper-branched β-glucan to provide various reaction sites for creating nanoparticles and their protection. Hyper-branched glucan can offer a variety of reaction sites for producing nanoparticles and protecting them because of its many structural variations. Hyperbranched-glucan has a significant hydroxyl group adsorption on the surface of the selenium that was used to manufacture water-dispersible selenium (Se) nanoparticles. It was 200 nm in size and stable in aqueous solution for 1 month (G. Liu et al., 2021). In a research mushroom polysaccharides were coated on Selenium nanoparticles, it handled the size of the particle was approx. 91–102 nm and prevented aggregation. Hyperbranched β-glucan has the property of preventing tumors without causing damage to the major organs, it is a better material as compared to other glucans (Li & Cheung, 2019).

Su et al. (2020) proved that to increase cellular absorption at a pH of 6.8, a pH-sensitive gene medication (CpG ODN) delivery system with hyperbranched -glucan coated with poly-ethylene glycol (PEG) shell was developed. This system allowed class B CpG-ODN which promoted multimerization. Therefore, hyperbranched β-glucan has more terminal hydroxyl groups that help in chemical and functional modifications and also the appropriate characterisation for creating the ideal target delivery system for pharmaceuticals and medicines is made possible by the vast surface area and shell structure.

### 7.3.4 pH-Dependent Delivery Systems

By proper entrainment of prebiotic polymers, the delivery systems could be improved and targeted for particular areas. Sorasitthiyanukarn et al. (2022) discovered that Astaxanthin is delivered orally using modified Ph-sensitive chitosan oligosaccharide-alginate nanoparticles (NPs), which improve the stability, qualities, and availability of astaxanthin. Upon storing and exposing to UV light, heat, and alkali environment the stability improved and at pH such as1.2, 6.8, 7.0, 7.4 and 37 °C changes and modifications were noted. In the first 6 h due to rapid diffusion, the sample burst released but with time the release was slow and within control. pH affected this in the manner that at low pH the release of the bioactive was faster. This was due to the large numbers of amino groups present which led to leakage otherwise at high pH of about 6.7, 7.0, and 7.4 there was shrinkage which led to slow release and entrapment of bioactive in the matrix. In 2018 using sodium alginate-cellulose nanofiber gel microsphere a pH-sensitive prebiotic carrier was developed. This carrier had capability for intestinal delivery of probiotics, alternate microbiome composition, and sheltered bacteria in gastric medium hence providing great applicability for delivery of targeted intestine (H. Zhang et al., 2018).

Another carrier known as a compartmentalised carrier is based on Chito oligosaccharide-ferritin nanocages and has a sphere-like form. It contains protective properties for carefully controlling the co-encapsulation of rutin and curcumin (H. Chen et al., 2022b). Oxidized konjac glucomannan (OKG)-based delivery carrier released the anthocyanin antioxidant in a controlled range. Due to faster degradation of carbohydrate microspheres in an acidic medium, the anthocyanin gets released in a large uncontrolled amount, this is prevented when there is an extra positive charge present on the carrier which prevents the entry of H+ ions and hence stops the early uncontrolled release of the antioxidant (Falsafi et al., 2022).

### *7.3.5 Time Dependent Delivery Systems*

These delivery carriers have the aim to utilize the time it takes the bioactive from ingestion to arriving at the targeted location. Various factors affect the transportation and delivery such as female having delayed gastric emptying and longer whole gut than males. These carriers focus on controlled and timely deliveries within the system nonetheless of the resistance and alterations that keep occurring in the stomach medium (Tsouris et al., 2014).

Oligofructose, inulin, and maltodextrin were used in a study to monitor the release of ascorbic acid and anthocyanins from camu-camu fruit at 25 and 35 degrees Celsius (de Abreu Figueiredo et al., 2020). Results showed that inulin and oligofructose microcapsules had higher hygroscopicity and density than maltodextrin carriers. Maltodextrin carrier had higher thermostability, antioxidant capability and bioactive retention but inulin and oligosaccharide based were more focused on controlling volatile entrapped bioactive extract (Misra et al., 2022). Pectin-non-starch like chitin, bacterial cellulose nanofibrous biocomposites was introduced as a gastrointestinal-resistant prebiotic carrier, it was responsible for survival rate for embedded *Bacillus coagulans* cells in difficult environments (Khorasani & Shojaosadati, 2017). Calcium ions were induced in the structure which enhanced the cross-linking of anionic units.

### *7.3.6 Colonic Microbiota Dependent Delivery Systems*

Once the bioactive are ingested orally they are dissolved in the gut. Delivery systems consisting of microbiota are advantageous to site-specific metabolic activity of microorganisms. Colon bacteria produce enzymes that can break cross-links between the bioactive chemical and carrier matrix and dissolve the coated matrix, releasing the payloads (Das, 2021). Upon reaching the colon, prebiotics are digested. When phenolic acids are grafted with oligosaccharides, their intestinal stability is increased, increasing resistance, and the gut microbiota then degrades them into free phenolic acids and oligosaccharides (D. Chen et al., 2022a). Antioxidant fabrication was done for *Lactobacillus casei* encapsulation conducted by Maillard reaction occurring between soy protein isolate and prebiotic oligosaccharides. The development of *L. casei* was promoted by this, which demonstrated remarkable stability against harmful microorganisms. In comparison to *L. casei* which was not treated with heat or stored, there was a greater percentage of survival under specific conditions. Another attempt evaluated the characteristics of Assam Bora rice's enzymatic stability in the upper digestive system and its degradability by colonic bacteria. Colon targeted bioactive delivery using starch-based microspheres. There is potential for colonic microbiota-dependent starch-based vehicles to be used as a strategy for targeted therapeutic delivery because, interestingly, just a small quantity of the starch was released in the upper GI tract following oral delivery, and the starch vehicles made it safely to the colon environment (92%) after oral administration (M. Z. Ahmad et al., 2013).

### 7.3.7 *Enzyme Sensitive Delivery Systems*

Several enzymes are present in colon such as β-D-glucosidase, β-D-galactosidase, amylase, dextranase, glycosidase, α-D-xylosidase, and pectinase, these can affect the glycosidic bond of polysaccharide and results in hydrolysation. In a study, whey protein isolate and flavonoids from yellow-skinned onions were combined with *Lactobacillus casei* to provide anti-diabetic and anti-inflammatory effects (Milea et al., 2020). A novel micrometer-sized chitosan carrier was designed for probiotic delivery (Table 7.2). Enzyme covering was provided for this to get proper delivery. This system had high stability and was degraded only by enzymes from colonic bacteria. The pH/time-dependent, enzyme-sensitive biopolymers and colonic microbiota-triggered systems continue to be among the most promising approaches for achieving tailored bioactive molecule delivery via oral route, despite the significant advancements in designing bioactive delivery systems using prebiotic-based carriers. Further in-depth research in this area is required since burst release and probable breakdown during GI transit, followed by the possible buildup, are all factors. Size range of from 10 to 300 m becomes a limiting factor in microparticulate prebiotic vehicles because bigger particles cause more particles to accumulate in the intestinal medium, which affects the absorption by cells (Moslemi, 2021).

## 7.4 Preparation Methods of β-Glucan-Based Delivery Systems

### 7.4.1 Yeast β-Glucan Based Delivery Systems

The yeast delivery system was constructed by using the cell wall β-glucan particles. Hollow β-glucan particles along with a layer of PEI and DNA covered the core of tRNA/polyethyleneimine (PEI) (Su et al., 2021). This structure is capable of acting as multifunctional carriers. A simple delivery system was created by incorporating β-glucan particles in a SiR-NA/EP compound made from amphipathic alpha-helical peptide Endo-porter (EP) system. Zymosan is reacted with trifluoracetic acid in DMSO (Table 7.2). This results in the formation of single strand fragments that create β-glucan nanoparticles bound by hydrogen bonding (Lima, 2020).

At a neutral pH in the presence of alkaline solution or dimethyl sulfoxide (DMSO) sometimes the β-1,3-glucan forms a single chain structure which gets converted into the original triple helical upon exchanging the solvent with water (Gidley & Nishinari, 2009). This also leads to formation of a complex with two β-1,3-glucan linkage and one polynucleotide such as poly (dA) attached by hydrogen bond and hydrophobic interactions. An appropriate amount of molecular weight is necessary for proper hydrogen bonding to create a stable structure. For single chain SPG (s-SPG) molecular weight of about 3000–5000 Da (16–30 glucose units) is required (Sakurai et al., 2005). SPG/dA-ODN needs 25,000 Da to form a complex of 50 nm with efficient delivery system through cell nucleus (Su et al., 2021). lateral chain Despite not taking part in the complex's creation, it is crucial to the oligonucleotide's and 1,3-glucan's ability to bind to extracellular receptors. Renatured triple helical structure SPG had less binding affinity than native structure whereas the SPG with phosphorothioate dA had more binding sites than native which led to enhanced electrostatic interactions (Mochizuki et al., 2014).

An investigation revealed that by adding arginine-glycine-aspartic acid tripeptide (RGD) at the side chain of SPG, antisense ODN absorption by cells was increased and this delivery method acquired a net negative charge (Matsumoto et al., 2004). This decreased the communication with serum protein to a bare minimum and became stabilized in the bloodstream. SPG reacted with carbon nanotubes, conductive polymers, gold/ silver nanoparticles within its helical structure and created various other novel structures and designs with better and enhanced stability, increased solubility and lesser cytotoxicity (Matiyani et al., 2021).

### 7.4.2 Hyper-Branched β-Glucan Based Delivery Systems

Hyperbranched delivery systems have different structures as compared to the normal structure, it has randomly arranged spherical or comb structures with various branching (Burchard, 1999).

## 7.5 Physiochemical Characterization of β-Glucan-Based Delivery System

Among the species of mushrooms *P. ostreatus* is one of the most important sources of β-glucan, about 32.3 g/100 g d.m. was found (Gallotti et al., 2020a).

### *7.5.1 Solubility*

Under particular circumstances, such as time and temperature, a polysaccharide's ability to disperse in a liquid medium and create a homogenous dispersion. Due to multiple hydroxyl groups polysaccharides have a strong attraction towards water molecules (Zeng et al., 2012). Solubility is affected by molecular structure and intermolecular aggregation, it is a vital factor because of its role in emulsification, film formation, drug delivery (Zargar et al., 2015). High-molecular weight β-glucan on dissolving in water forms a microgel like structure and enhances the cholesterol lowering properties. The structure was well defined and highly aggregated, it was modified chemically by sulfation process, gamma irradiation or ultrasonic treatment resulting in an increase in solubility and a decrease in molecular weight (Y. Zhang et al., 2021). Particle size and water content affect the solubility of β-glucan, smaller particle has greater solubility, and less water presence in the coating makes the structure harder and less soluble (Pieczykolan & Kurek, 2019).

### *7.5.2 Viscosity*

The polysaccharide's viscosity is influenced by concentration, molar mass, and the cap's ability to form aggregates. A study revealed that whereas oat fibre had high viscosity at lower shear rates, β-glucan had low molar mass, low viscosity, and high solubility (Lambo et al., 2005). Oat β-glucan (OBG) with β-(1,4)-to β-(1,3)-bond had high viscosity whereas the barley β-glucan (BBG) had low viscosity (Mikkelsen et al., 2010).

### *7.5.3 Gelation*

Gelling property of β-glucan varies according to the molecular weight, components with high molecular weight took more time to form gel than low molecular weight components (Tosh et al., 2004). Hydrogen peroxide/ ascorbic acid treatment degrades polymers which results in forming weaker gels for the oxidized BG. This was useful for regulating the gelling properties and other features making it an efficient and economical method for modifying BG and improving the use in food industry (Khorami et al., 2014).

### 7.5.4 Binding Capacity

Enzyme such as cholesterol 7α-hydroxylase converts cholesterol into bile acid in liver which gets reabsorbed via active and passive mechanism through enterohepatic circulation. Bile acid production rises if bile acids are physically removed from the enterohepatic circulation. This can be accomplished by inhibiting ileal reabsorption, which ultimately leads to an increase in faecal excretion (Gonzalez, 2012). When bile acid is consumed and reaches the small intestine, OBG interacts with it and lowers the blood cholesterol. This is affected by chemical modification such as sulfation reduces the ability to bind with bile acid whereas acetylation increases the ability (Song, 2014).

There are two ways in which plant compounds interact with bile acid, first is direct molecular connection with each other and the second mechanism is that with an increase in viscosity of small intestine materials there will be decrease in fluidity of micelles (Koziolek et al., 2019). Removing stool from bile acid reduces amount of recirculation resulting in more cholesterol formation. The U.S. Food Administration (FDA) declared that oat bran and BG can be used for cholesterol lowering food which also helps in eventually decreasing heart disease risk (Goldberg, 1992).

### 7.5.5 Bulk Density

Different microcapsules have different densities, highest density on observing turned out to be of inulin (IN) capsules (0.95 g/cm$^3$) and lowest was that of gum Arabic (GA) (0.78 g/cm$^3$) (Benjasirimongkol et al., 2019). Different values are distinguished by the relationship between molecular weight of the coating material and the bulk density. Less bulk density is due to the components having less mass which is a task to fill spaces between molecules, hence occupying more space. Bulk density is also affected by other parameters such as shape, size, surface area, drying speed (Ferrari et al., 2012). In quick drying the component retains its original shape by hardening but in slow drying the shape converts into concave after evaporation (Lowinger et al., 2014).

### 7.5.6 Efficiency of Microencapsulation

It is one of the most important factors for determining the capacity of the wall material to encapsulate the core material inside the microcapsule, high retention of the core material with minimum surface is considered as an efficient method. It is affected by the properties of the wall and core materials. Different coating materials were compared in the efficiency of microencapsulation, results showed GA had the lowest efficiency of 78.61% and guar gum (GG) had the highest of 92.98% and IN had the efficiency of 88.37% (Pieczykolan & Kurek, 2019). Lowest efficiency was shown by the material commonly used for encapsulation.

### 7.5.7 Anthocyanins and Ascorbic Acid Content

Anthocyanin content was determined, highest was of BG capsules (3055.65 mg/100 g) and lowest was of GA (1940.22 mg/100 g) (Y.-R. Lee et al., 2009). Hollow capsule could be created due to the high solubility of the coating material, it allows the entrapment of the extract in the crust. (Villacrez et al., 2014) Discovered that anthocyanin didn't affect the storage of the core material but on the other hand Hongmei and Meng calculated that storage decreased to half life of anthocyanin at room temperature than in refrigerator.

Calculating total ascorbic acid revealed that PEC had the highest value (12.19), demonstrating that pectin has a protective effect with respect to vitamin C. the terms of percentage, the scored the most (47%) while GG came the last (8%). GG deteriorated anthocyanins during storage out of all the coating materials (Eshghi et al., 2022).

As we already have discussed β-glucans are polysaccharide found in cereals, fungi, and bacteria with different types of linkage and variability in the structure. Simplest are found in cell wall or source for energy storage for example curdlan with the β-(1,3) linear structure produced by gram negative bacteria *Agrobacterium* is used as a food additive and stabilizer. Physico-chemical and biocompatible properties of β-glucan allow it to act as a coating material for nanoparticles which prevents the issue of cytotoxicity and aggregation. β-(1,3; 1,6)-D-glucan derived from *Pleurotus florida*, and edible mushroom created stable gold nanoparticles by reducing chloroauric acid. Stable silver nanoparticles were derived from carboxymethyl curdlan to reduce nitrates (Li & Cheung, 2019). By using hyperbranched β-glucan water dispersible selenium nanoparticles were formed, they had better stability and water absorption capacity. Apart from these other nanoparticles with stable and cost-effective properties have been developed that can provide various opportunities for biopharmaceutical industries. We have seen β-glucan consist of various health related properties such as anti-tumour, antiviral, immunostimulant activity, along with β-glucan purified from alkali-soluble polysaccharide plays a role in protecting biopharmaceuticals and providing reaction sites for chemical modifications. This makes the β-glucan isolated from *P. tuberregium* (HBB Ptr) a simple and effective component for creating a delivery system for the encapsulation of biopharmaceuticals. HBB Ptr enhances the stability of the nanoparticles making them beneficial for cancer therapy with no obvious side effects.

In terms of safety of β-glucan based delivery systems, an experiment was conducted using barley β-glucan (Glucagel) on lab rats which resulted in no toxicity even after consuming high concentrations of glucan. Glucagel was extracted from barley using water-based process, it was stable at room temperature. Four groups of five rats per sex were taken with 0%, 1%, 5% or 10% Glucagel concentrations for 28 days. Behavioural changes, clinical changes, physical changes were observed regularly to check the effects of Glucagel on the animal. Results were analysed and the level of β-glucan was near to the expected values before and after the time period hence displaying stability of the glucan. No disease occurred in the particular time period and all the rats were alive and well after 28 days of these experiments therefore no treatment was required. The average daily consumption of Glucagel was 7.7 g/kg for men and 7.8 g/kg for women, delivering 5.8–5.9 g of β-glucan per kilogramme of body weight per day. The body weight was nearly same before and after (Delaney et al., 2003). All of this represented that there was no adverse or harmful effect. In high dosage male rats, the plasma of cholesterol and phospholipids concentrations were less, and no toxicity was observed, there was high plasma level of urea but not in females hence no toxicological significance was found. Only increases in full and empty caecum weight were observed which was due to consuming high amount of indigestible carbs and had no toxic effects. In conclusion, consumption of food with up to 10% Glucagel for 28 days did not have any toxicity. This effect of β-glucan in rats above the expected human intake without any toxic effects supports the claim that Glucagel was safe to use under specific circumstances. This consumption of β-glucan was 100-fold more than that recommended for decreasing the blood cholesterol level, i.e., 3 g/person/day, 0.05 g/kg body weight/day for a 60 kg weighing person (Jonker et al., 2010).

Another experiment was conducted using "yeast β-glucan", market "yeast β-glucan" was taken and induced in food material at a level of 375 mg/day and 600 mg/day in particular nutritional food materials (Agostoni et al., 2011). These were put in different types and varieties of foods including beverages but not food for infants and formula. Results showed that intake of high level of "yeast β-glucan" had the same effects as ingesting β-glucan from any other dietary source, it had no toxic effects, no disease occurred and there was no point of concern. EFSA (European Food Safety Authority) has already assessed the safety of beta-glucans from various sources. Given the characteristics of "yeast beta-glucans," the extensive history of use of its source, the estimated intake provided, and further data from research on both humans and animals, "yeast beta-glucans" is safe at the doses now being used.

## 7.6 In Vitro and In Vivo Evaluation of β-Glucan Based Delivery Systems

Monomeric anthocyanins were removed efficiently in vitro conditions through the coating of the wall material, it was observed that the anthocyanins released from saffron anthocyanins β-glucan ($S_{\beta}G$) were low in amount as compared to saffron anthocyanins in β-cyclodextrin ($S_{\beta}C$) as mentioned in Table 7.3. The covering is

**Table 7.3** *In vitro* digestion of anthocyanin from saffron based encapsulated powders

| Digestion | Duration (min) | $S_{\beta}C^{a}$ (%) | $S_{\beta}G^{a}$ (%) | References |
|---|---|---|---|---|
| Mouth | 5 | 15.83 | 11.50 | (M. Ahmad et al., 2018) |
| Gastric conditions | 30 | 38.63 | 22.93 | (M. Ahmad et al., 2018) |
| | 60 | 41.26 | 39.83 | |
| Intestinal conditions | 120 | 51.56 | 61.56 | (M. Ahmad et al., 2018) |
| | 240 | 71.76 | 80.33 | |

[a]$S_{\beta}C$ and $S_{\beta}G$ are encapsulated anthocyanins from saffron in β glucan and β-cyclodextrin, respectively

damaged and microparticles leak out anthocyanins as a result of the stomach's neutral pH. This study showed that particles which were coated with β-glucan and β-cyclodextrin displayed resistance against gastric environment of stomach and protected the bioactive compounds. These were separated and transferred to simulated intestinal fluid and maximum anthocyanins were released at the first 2 h, with advancement in time there was enhancement in the absorption of bioactive compounds (M. Ahmad et al., 2018). Another experiment was conducted in vivo conditions to acknowledge the immunity increasing capability of β-glucan. The rate of protection in the β-group vaccine and anisodamine, chitosan group vaccine was termed as higher as compared to the control group vaccine. This suggested that inactivated vaccine had a greater ability to prevent infection from CyHV-2 (cyprinid herpesvirus 2) and that -glucan strengthened the protective properties of the vaccination. A test for antibody neutralisation under in vitro conditions was performed to determine the IgM concentrations in gibel carp serum. Serum IgM levels in the vaccination and glucan vaccine groups were elevated, and when serially diluted sera were combined with CyHV-2 and gibel carp (GiCB) cells were further incubated for neutralisation, the significance of the glucan vaccine treatment was raised (Huo et al., 2024).

In recent years there have been studies about the property of β-glucan of binding bile acid, in vitro conditions the interaction between β-glucan and bile acid was examined giving results that chemical modification such as sulfation caused a decrease in the binding but on the other hand acetylation process enhanced the binding (Kim & White, 2011). In vivo terms fecal bile acid discharge in higher amounts along with stool release decreases recirculation and leads to the formation of cholesterol in the liver. Due to this, a gel is formed by β-glucan in the small intestine which improves the reabsorption and reduces the amount of cholesterol henceforth making β-glucan consumable at 3 g/day to reduce cholesterol and heart disease risk (Rostoll Cangiano, 2023). α-tocopherol due to its varying swelling properties is released from the formulated model system under certain conditions. The findings indicated that while β-tocopherol was released during 2 h, the amounts of β-tocopherol from β-glucan and β-glucan nanoparticle (Glu-N) fluctuated. Under simulated gastric conditions, Glu and Glu-N generated 14.8 and 71.2 g/mL of β-tocopherol in the first 30 min of digestion, which varied significantly (*p* 0.05). Next 60 min. The release of 120 and 25.2 g/mL of -tocopherol by Glu-N and Glu during the 60 min again showed a substantial difference between the two, demonstrating that it will swiftly reach the intestines for absorption. Therefore, Glu gave a delayed release under SG circumstances. Once more, under artificial intestinal conditions, Glu-N delivered a greater release of -tocopherol than Glu did at intervals of 30 and 60 min. When bile was digested for 120 min, slow releases of Glu (40 g/mL) and Glu-N (187 g/mL) were produced (ul Ashraf et al., 2020).

In vitro tests on the human breast adenocarcinoma cell line (MCF-7), human ductal breast epithelial tumour cell line (T47D), human colon cancer cell line (Colo-25), SW480, and HT-29 revealed nanohybrids to be non-toxic to normal cells and to produce cell ablation under heat. The same in vivo experiments caused reduction in the solid tumours through consuming by intravenous injection consisting 25 mg/kg body weight (Abed et al., 2022). The gold nanorods did not contain any toxicity due to the hyper-branched β-glucan in vitro and in vivo because the chemical cetylrimethylammonium bromide (CTAB) blocked the surface and produced nano-hybrids. The hyperbranched structure also had antitumor properties without damaging any major organs (Su, 2021). Antimicrobial activities of β-glucan against the gram-positive Bacillus subtilis and the gram-negative *E. coli* bacteria were demonstrated in vitro. It is possible to connect the size of β-glucan nanoparticles with their improved antibacterial activity. In comparison to bigger particles, smaller particles have a greater surface area and a higher likelihood of interacting with bacterial cell membranes. Due to the increased ease with which -glucan nanoparticles may enter and harm the bacterial cell, a larger inhibition zone results. The results also indicated that gram-positive *B. subtilis* is more resistant to β-glucan nanoparticles from various sources than gram-negative *E. coli*. Because both the surfaces of the nanoparticles and the bacteria have negative potentials, the bulk of interactions at the contact between them are electrostatically repellent. The nanoparticles do not adhere to the bacterial surface as a result (Mocanu et al., 2019).

## 7.7 Applications of β-Glucan-Based Delivery systems in Drug Delivery: Micro and Nanoencapsulation Technologies

The potential of β-D-glucan to influence the immune system may make it valuable as a target for immunotherapy due to its capacity to bind to specific receptors on immune cells including macrophages and monocytes. A delivery approach that targets macrophages and has potent cellular absorption through the receptor Dectin-1 has utilised several β-D-glucans. More and more nano-carriers are being created that can stop medications or genes from degrading while in circulation. To target cancer cells in response to physiological cues such as changes in pH, redox potential, and enzyme concentration, pH-responsive and glutathione (GSH)-triggered release systems have also been created (Such et al., 2015). They tend to promote the growth of macrophages because β-D-glucans increase GM-CSF via Dectin-1-independent ERK and p38 MAPK activation, which contributes to their immunotherapeutic potential. A fungal β-D-glucan may alter monocyte metabolism by activating the Dectin-1-Akt-mTOR-HIF-1 (hypoxia-inducible factor-1) pathway, leading to increased aerobic glycolysis. By switching from oxidative phosphorylation to aerobic glycolysis, activated macrophages, dendritic cells, and T lymphocytes may improve the immune system's response to infection and inhibit the formation of malignancies (Batbayar et al., 2012). It has been suggested that yeast-derived β-D-glucan may be able to transform immune-suppressive TAMs into immune-stimulating M1 phenotype macrophages. This is due to changes in macrophage metabolism, including enhanced glycolysis, Krebs cycle activity, and glutamine utilisation, which is mediated by Dectin-1. A new anticancer vaccine built from a synthetic MUC1 peptide and β-D-glucan suggests that β-D-glucan can function as a potential transporter and immune activator for increasing the stability and immunogenicity of peptide antigens (Su et al., 2021). It was claimed that the vaccination helped with antigen transport, immune cell identification, and activation. The β-glucan PAMP (Pathogen Associated Molecular Pattern), which was created in conjunction with anti-PD-1/PD-L1, demonstrated improved antitumor efficacy in syngeneic tumour models with increased antigen presenting capacity, repolarization of M2 macrophages, and maturation of dendritic cells in a clinical trial. It is critical to stress that using β-D-glucan as a carrier or in combination with different immunotherapeutic methods is receiving increasing attention (Fraser et al., 2016).

## 7.8 Future Directions and Challenges in β-Glucan-Based Drug Delivery

The properties of β-glucan make it possible for it to be applied in various industries and fields with magnificent results, in the future β-glucan has a great scope in preparing vaccines, stimulating immunity, drug carriers and transporters and also could be desirable as feed for animals. However, there are still problems and challenges that β-glucan will face such as on the one hand, β-glucans produced using various extraction, separation, and purification techniques exhibited varying degrees of purity, molecular weights, and spatial structures, which resulted in varying biological activity. On the other hand, primarily β-glucan enhanced the animal's capacity to combat harmful microorganisms and stress, and as a result, they had varying impacts on diverse animals used in the production process. The appropriate structure-function connection of β-glucan is currently problematic. The active mechanism, spatial organisation, molecular weight, and linkage of β-glucan require more study. The biological activities of glucan have also been significantly boosted by the present modification strategies, although the physical change had the disadvantage of being an unpredictable reaction with an unknown molecular weight range. If we focus on the side chain structure of β-glucan, we can find that certain of the groups connected by the side chains exhibit extremely high biological activity (Sikora et al., 2013). The molecular weight range with the best biological activities was found by investigating the direct relationship between polysaccharide molecular weight and biological activities. The stable and efficient modification strategies for β-glucan were determined by comparing the currently used β-glucan modification methods. Studying derivatives' unique biological activities, their structures, and the effects of the structure-activity and dose-activity connections on the biological activities were all key steps in furthering our understanding of biology. The biological activities of β-glucan may be greatly increased through research, increasing its accessibility for animal production (Rabah et al., 2022).

To improve β-D-glucans' specific interaction with target cells and create more efficient delivery systems with high specificity, screening will be required in the future. However, while certain treatment medications are designed to reduce the number of macrophages, it is still difficult to reduce the detrimental effects on healthy tissue-specific macrophages that are essential for sustaining bodily function. A sensitive detachable covering made using nanotechnology that might protect internal organs while they circulate in the bloodstream but release them at the desired location in reaction to physiological cues like pH, redox potential, or enzyme concentration changes in the tumor microenvironment (TME) could be an answer. Such a strategy can balance the delivery system's stability with precise targeting. Accordingly, a promising option for cancer immunotherapy would be a delivery system based on β-D-glucan that focuses on Targeting tumor-associated macrophages (TAMs) and uses nanotechnology in conjunction with a TAMs-targeted strategy. Through their very specific imaging, such a delivery mechanism would also aid future studies on TAMs as a predictive indicator of cancer.

## 7.9 Conclusion

β-Glucan is a promising bioactive carrier, providing stability, regulated release, and health advantages. Its ability to encapsulate sensitive components such as rutin and tocopherols efficiently supports its use in functional foods and pharmaceuticals. Nanoencapsulation with β-glucan matrices enhances bioavailability and protects bioactives under simulated gastrointestinal conditions. Additionally, β-glucan-based systems have prebiotic qualities, improve antibacterial and antioxidant activity, and distribute well across the gastrointestinal tract. Despite its enormous promise, obstacles remain in understanding the structure-function link, optimizing its physicochemical qualities, and ensuring its use in a variety of food and medicinal systems. Further study should include clinical validation, scaling up nanoencapsulation techniques, and investigating β-glucan's synergy with alternative delivery systems. By resolving these problems, β-glucan has the potential to improve health outcomes and product innovation in nutraceuticals and other fields.

## References

Abdel-Haleem, A. M. H., & Awad, R. A. (2015). Some quality attributes of low fat ice cream substituted with hulless barley flour and barley ß-glucan. *Journal of Food Science and Technology, 52*, 6425–6434.

Abed, A., Derakhshan, M., Karimi, M., Shirazinia, M., Mahjoubin-Tehran, M., Homayonfal, M., Hamblin, M. R., Mirzaei, S. A., Soleimanpour, H., Dehghani, S., et al. (2022). Platinum nanoparticles in biomedicine: Preparation, anti-cancer activity, and drug delivery vehicles. *Frontiers in Pharmacology, 13*, 797804.

Agostoni, C. V., Bresson, J.-L., Fairweather-Tait, S., Flynn, A., Golly, I., Korhonen, H., Lagiou, P., Lovik, M., Marchelli, R., Martin, A., et al. (2011). Scientific opinion on the safety of'yeast beta-glucans' as a novel food ingredient. *EFSA Journal, 9*(5).

Ahmad, M., Ashraf, B., Gani, A., & Gani, A. (2018). Microencapsulation of saffron anthocyanins using β glucan and β cyclodextrin: Microcapsule characterization, release behaviour & antioxidant potential during in-vitro digestion. *International Journal of Biological Macromolecules, 109*, 435–442.

Ahmad, M. Z., Akhter, S., Dhiman, I., Sharma, P., & Verma, R. (2013). Evaluation of material properties and compression characteristics of Assam Bora rice flours as a directly compressible vehicle in tablet formulation. *Expert Opinion on Drug Delivery, 10*(2), 163–171.

Akagi, T., Baba, M., & Akashi, M. (2012). Biodegradable nanoparticles as vaccine adjuvants and delivery systems: Regulation of immune responses by nanoparticle-based vaccine. In *Polymers in nanomedicine* (pp. 31–64).

Batbayar, S., Lee, D. H., & Kim, H. W. (2012). Immunomodulation of fungal β-glucan in host defense signaling by dectin-1. *Biomolecules & Therapeutics, 20*(5), 433.

Benjasirimongkol, P., Sriamornsak, P., et al. (2019). *Development and characterization of resveratrol dry emulsions using Quality-by-Design approach*. Silpakorn University.

Bi, H., Xu, Y., Fan, F., & Sun, X. (2022). Effect of drying methods on Lactobacillus Rhamnosus GG microcapsules prepared using the complex coacervation method. *Journal of Food Science, 87*(3), 1282–1291.

Breedveld, M. W., & Miller, K. J. (1994). Cyclic beta-glucans of members of the family Rhizobiaceae. *Microbiological Reviews, 58*(2), 145–161.

Burchard, W. (1999). Solution properties of branched macromolecules. In *Branched polymers II* (pp. 113–194). Springer.

Chen, D., Bai, R., Yong, H., Zong, S., Jin, C., & Liu, J. (2022a). Improving the digestive stability and prebiotic effect of carboxymethyl chitosan by grafting with gallic acid: In vitro gastrointestinal digestion and colonic fermentation evaluation. *International Journal of Biological Macromolecules, 214*, 685–696.

Chen, H., Han, X., Fu, Y., Dai, H., Wang, H., Ma, L., & Zhang, Y. (2022b). Compartmentalized chitooligosaccharide/ferritin particles for controlled co-encapsulation of curcumin and rutin. *Carbohydrate Polymers, 290*, 119484.

Chen, H., Liu, N., He, F., Liu, Q., & Xu, X. (2022c). Specific β-glucans in chain conformations and their biological functions. *Polymer Journal, 54*(4), 427–453.

Chillo, S., Ranawana, D. V., Pratt, M., & Henry, C. J. K. (2011). Glycemic response and glycemic index of semolina spaghetti enriched with barley β-glucan. *Nutrition, 27*(6), 653–658.

Das, S. (2021). Pectin based multi-particulate carriers for colon-specific delivery of therapeutic agents. *International Journal of Pharmaceutics, 605*, 120814.

de Abreu Figueiredo, J., Teixeira, M. A., Campelo, P. H., Lago, A. M. T., de Souza, T. P., Yoshida, M. I., de Oliveira, C. R., Pereira, A. P. A., Pastore, G. M., Sanches, E. A., et al. (2020). Encapsulation of camu-camu extracts using prebiotic biopolymers: Controlled release of bioactive compounds and effect on their physicochemical and thermal properties. *Food Research International, 137*, 109563.

Delaney, B., Carlson, T., Frazer, S., Zheng, T., Hess, R., Ostergren, K., Kierzek, K., Haworth, J., Knutson, N., Junker, K., et al. (2003). Evaluation of the toxicity of concentrated barley β-glucan in a 28-day feeding study in Wistar rats. *Food and Chemical Toxicology, 41*(4), 477–487.

Desai, N., Rana, D., Salave, S., Gupta, R., Patel, P., Karunakaran, B., Sharma, A., Giri, J., Benival, D., & Kommineni, N. (2023). Chitosan: A potential biopolymer in drug delivery and biomedical applications. *Pharmaceutics, 15*(4), 1313.

Drozińska, E., Kanclerz, A., & Kurek, M. A. (2019). Microencapsulation of sea buckthorn oil with β-glucan from barley as coating material. *International Journal of Biological Macromolecules, 131*, 1014–1020. https://doi.org/10.1016/j.ijbiomac.2019.03.150

Emmons, C. L., & Peterson, D. M. (1999). Antioxidant activity and phenolic contents of oat groats and hulls. *Cereal Chemistry, 76*(6), 902–906.

Eshghi, S., Karimi, R., Shiri, A., Karami, M., & Moradi, M. (2022). Effects of polysaccharide-based coatings on postharvest storage life of grape: Measuring the changes in nutritional, antioxidant and phenolic compounds. *Journal of Food Measurement and Characterization, 16*(2), 1159–1170.

Falsafi, S. R., Bangar, S. P., Chaudhary, V., Hosseini, E., Mokhtari, Z., Karaca, A. C., Samota, M. K., Goswami, D., Krishnan, V., Askari, G., et al. (2022). Recent advances in oral delivery of bioactive molecules: Focus on prebiotic carbohydrates as vehicle matrices. *Carbohydrate Polymers, 298*, 120074.

Ferrari, C. C., Germer, S. P. M., Alvim, I. D., Vissotto, F. Z., & de Aguirre, J. M. (2012). Influence of carrier agents on the physicochemical properties of blackberry powder produced by spray drying. *International Journal of Food Science & Technology, 47*(6), 1237–1245.

Filipczak, N., Pan, J., Yalamarty, S. S. K., & Torchilin, V. P. (2020). Recent advancements in liposome technology. *Advanced Drug Delivery Reviews, 156*, 4–22.

Fraj, A., Jaâfar, F., Marti, M., Coderch, L., & Ladhari, N. (2019). A comparative study of oregano (Origanum vulgare L.) essential oil-based polycaprolactone nanocapsules/microspheres: Preparation, physicochemical characterization, and storage stability. *Industrial Crops and Products, 140*, 111669.

Fraser, K., Chan, A., Fulton, R., Leonardo, S., Jonas, A., Qiu, X., Ottoson, N., Kangas, T., Gordon, K., Graff, J., et al. (2016). Imprime PGG triggers PD-L1 expression on tumor and myeloid cells and prevents tumor establishment in combination with αPD-L1 treatment in vivo. *Cancer Research, 76*(14_Suppl), 2335.

Gallotti, F., Lavelli, V., & Turchiuli, C. (2020a). Application of Pleurotus ostreatus β-glucans for oil–in–water emulsions encapsulation in powder. *Food Hydrocolloids, 105*, 105841.

Gallotti, F., Lavoisier, A., Turchiuli, C., & Lavelli, V. (2020b). Impact of pleurotus ostreatus β-glucans on oxidative stability of active compounds encapsulated in powders during storage and in vitro digestion. *Antioxidants, 9*(12), 1–16. https://doi.org/10.3390/antiox9121219

Gidley, M. J., & Nishinari, K. (2009). Physico-chemistry of (1, 3)-β-glucans. In *Chemistry, biochemistry, and biology of 1–3 beta glucans and related polysaccharides* (pp. 47–118). Elsevier.

Goldberg, J. P. (1992). Nutrition and health communication: The message and the media over half a century. *Nutrition Reviews, 50*(3), 71–77.

Gonzalez, F. J. (2012). Nuclear receptor controlof enterohepatic circulation. *Comprehensive Physiology, 2*(4), 2811–2828.

Han, B., Baruah, K., Cox, E., Vanrompay, D., & Bossier, P. (2020). Structure-functional activity relationship of β-glucans from the perspective of immunomodulation: A mini-review. *Frontiers in Immunology, 11*, 658.

Hennink, W. E., & van Nostrum, C. F. (2012). Novel crosslinking methods to design hydrogels. *Advanced Drug Delivery Reviews, 64*, 223–236.

Houacine, C., Yousaf, S. S., Khan, I., Khurana, R. K., & Singh, K. K. (2018). Potential of natural biomaterials in nano-scale drug delivery. *Current Pharmaceutical Design, 24*(43), 5188–5206.

Huo, X., Yan, Y., Chang, J., & Su, J. (2024). Astragalus polysaccharide or β-glucan combined with inactivated vaccine markedly prevent CyHV-2 infection in *Carassius auratus gibelio*. *Aquaculture and Fisheries, 9*, 763–775.

Islam, M. Z., An, H.-G., Kang, S.-J., & Lee, Y.-T. (2021). Physicochemical and bioactive properties of a high β-glucan barley variety 'Betaone'affected by germination processing. *International Journal of Biological Macromolecules, 177*, 129–134.

Jonker, D., Hasselwander, O., Tervilä-Wilo, A., & Tenning, P. P. (2010). 28-Day oral toxicity study in rats with high purity barley beta-glucan (Glucagel™). *Food and Chemical Toxicology, 48*(1), 422–428.

Khorami, H. A., Botero-Cadavid, J. F., Wild, P., & Djilali, N. (2014). Spectroscopic detection of hydrogen peroxide with an optical fiber probe using chemically deposited Prussian blue. *Electrochimica Acta, 115*, 416–424.

Khorasani, A. C., & Shojaosadati, S. A. (2017). Pectin-non-starch nanofibers biocomposites as novel gastrointestinal-resistant prebiotics. *International Journal of Biological Macromolecules, 94*, 131–144.

Kim, H. J., & White, P. J. (2011). Optimizing the molecular weight of oat β-glucan for in vitro bile acid binding and fermentation. *Journal of Agricultural and Food Chemistry, 59*(18), 10322–10328.

Koziolek, M., Alcaro, S., Augustijns, P., Basit, A. W., Grimm, M., Hens, B., Hoad, C. L., Jedamzik, P., Madla, C. M., Maliepaard, M., et al. (2019). The mechanisms of pharmacokinetic food-drug interactions—A perspective from the UNGAP group. *European Journal of Pharmaceutical Sciences, 134*, 31–59.

Kumar, A., Sharipov, M., Turaev, A., Azizov, S., Azizov, I., Makhado, E., Rahdar, A., Kumar, D., & Pandey, S. (2022). Polymer-based hybrid nanoarchitectures for cancer therapy applications. *Polymers, 14*(15), 3027.

Lambo, A. M., Öste, R., & Nyman, M. E.-L. (2005). Dietary fibre in fermented oat and barley β-glucan rich concentrates. *Food Chemistry, 89*(2), 283–293.

Lee, D. H., & Kim, H. W. (2014). Innate immunity induced by fungal β-glucans via dectin-1 signaling pathway. *International Journal of Medicinal Mushrooms, 16*(1), 1–16.

Lee, Y.-R., Nho, J.-W., Hwang, I.-G., Kim, W.-J., Lee, Y.-J., & Jeong, H.-S. (2009). Chemical composition and antioxidant activity of ramie leaf (Boehmeria nivea L.). *Food Science and Biotechnology, 18*(5), 1096–1099.

Li, X., & Cheung, P. C. K. (2019). Application of natural β-glucans as biocompatible functional nanomaterials. *Food Science and Human Wellness, 8*(4), 315–319.

Lima, T. I. P. (2020). *Modulation of the host immune response to a human fungal pathogen using β-glucan-functionalized nanoparticles.*

Liu, C., Choi, M. W., Li, X., & Cheung, P. C. K. (2018a). Immunomodulatory effect of structurally-characterized mushroom sclerotial polysaccharides isolated from Polyporus rhinocerus on human monoctyes THP-1. *Journal of Functional Foods, 41*, 90–99.

Liu, G., Yang, X., Zhang, J., Liang, L., Miao, F., Ji, T., Ye, Z., Chu, M., Ren, J., & Xu, X. (2021). Synthesis, stability and anti-fatigue activity of selenium nanoparticles stabilized by Lycium barbarum polysaccharides. *International Journal of Biological Macromolecules, 179*, 418–428.

Liu, M., Wu, T., Li, S., Wei, P., Yan, Y., Gu, W., Wang, W., & Meng, Q. (2018b). Combined transcriptomic/proteomic analysis of crucian carp Carassius auratus gibelio in cyprinid herpesvirus 2 infection. *Fish & Shellfish Immunology, 82*, 386–399.

Liu, Q., Duan, B., Xu, X., & Zhang, L. (2017). Progress in rigid polysaccharide-based nanocomposites with therapeutic functions. *Journal of Materials Chemistry B, 5*(29), 5690–5713.

López-Rubio, A., & Lagaron, J. M. (2012). Whey protein capsules obtained through electrospraying for the encapsulation of bioactives. *Innovative Food Science & Emerging Technologies, 13*, 200–206.

Lowinger, M., Baumann, J., Vodak, D. T., & Moser, J. (2014). Practical considerations for spray dried formulation and process development. In *Discovering and developing molecules with optimal drug-like properties* (pp. 383–435). Springer.

Matiyani, M., Pathak, M., Bohra, B. S., & Sahoo, N. G. (2021). Noncovalent functionalization of carbon nanotubes. In *Handbook of carbon nanotubes* (pp. 1–28). Springer.

Matsumoto, T., Numata, M., Anada, T., Mizu, M., Koumoto, K., Sakurai, K., Nagasaki, T., & Shinkai, S. (2004). Chemically modified polysaccharide schizophyllan for antisense oligonucleotides delivery to enhance the cellular uptake efficiency. *Biochimica et Biophysica Acta (BBA)-General Subjects, 1670*(2), 91–104.

Mikkelsen, M. S., Jespersen, B. M., Møller, B. L., Lærke, H. N., Larsen, F. H., & Engelsen, S. B. (2010). Comparative spectroscopic and rheological studies on crude and purified soluble barley and oat β-glucan preparations. *Food Research International, 43*(10), 2417–2424.

Milea, Ş. A., Vasile, M. A., Crăciunescu, O., Prelipcean, A.-M., Bahrim, G. E., Râpeanu, G., Oancea, A., & Stănciuc, N. (2020). Co-microencapsulation of flavonoids from yellow onion skins and lactic acid bacteria lead to multifunctional ingredient for nutraceutical and pharmaceutics applications. *Pharmaceutics, 12*(11), 1053.

Misra, S., Pandey, P., Dalbhagat, C. G., & Mishra, H. N. (2022). Emerging technologies and coating materials for improved probiotication in food products: A review. *Food and Bioprocess Technology, 15*(5), 998–1039.

Mocanu, A., Isopencu, G., Busuioc, C., Popa, O.-M., Dietrich, P., & Socaciu-Siebert, L. (2019). Bacterial cellulose films with ZnO nanoparticles and propolis extracts: Synergistic antimicrobial effect. *Scientific Reports, 9*(1), 1–10.

Mochizuki, S., Morishita, H., Adachi, Y., Yamaguchi, Y., & Sakurai, K. (2014). Binding assay between murine Dectin-1 and β-glucan/DNA complex with quartz-crystal microbalance. *Carbohydrate Research, 391*, 1–8.

Mochizuki, S., & Sakurai, K. (2009). A novel polysaccharide/polynucleotide complex and its application to bio-functional DNA delivery system. *Polymer Journal, 41*(5), 343–353.

Moslemi, M. (2021). Reviewing the recent advances in application of pectin for technical and health promotion purposes: From laboratory to market. *Carbohydrate Polymers, 254*, 117324.

Murphy, E. J., Rezoagli, E., Major, I., Rowan, N. J., & Laffey, J. G. (2020). β-glucan metabolic and immunomodulatory properties and potential for clinical application. *Journal of Fungi, 6*(4), 356.

Neri-Numa, I. A., Arruda, H. S., Geraldi, M. V., Júnior, M. R. M., & Pastore, G. M. (2020). Natural prebiotic carbohydrates, carotenoids and flavonoids as ingredients in food systems. *Current Opinion in Food Science, 33*, 98–107.

Ng, H.-M., Sin, L. T., Tee, T.-T., Bee, S.-T., Hui, D., Low, C.-Y., & Rahmat, A. R. (2015). Extraction of cellulose nanocrystals from plant sources for application as reinforcing agent in polymers. *Composites Part B: Engineering, 75*, 176–200.

Nishantha, M. D. L. C., Zhao, X., Jeewani, D. C., Bian, J., Nie, X., & Weining, S. (2018). Direct comparison of β-glucan content in wild and cultivated barley. *International Journal of Food Properties, 21*(1), 2218–2228.

Ozcan, T., Akpinar-Bayizit, A., & Yilmaz-Ersan, L. (2017). The functional aspects of beta glucan for dairy industry. *Science within Food: Up-to-date Advances on Research and. Educational Ideas, 14*, 264.

Pateiro, M., Gómez, B., Munekata, P. E. S., Barba, F. J., Putnik, P., Kovačević, D. B., & Lorenzo, J. M. (2021). Nanoencapsulation of promising bioactive compounds to improve their absorption, stability, functionality and the appearance of the final food products. *Molecules, 26*(6), 1547.

Pieczykolan, E., & Kurek, M. A. (2019). Use of guar gum, gum arabic, pectin, beta-glucan and inulin for microencapsulation of anthocyanins from chokeberry. *International Journal of Biological Macromolecules, 129*, 665–671.

Qureshi, A., et al. (2023). *The effect of germination on the physicochemical, functional, and nutritional properties of yellow pea, red lentil and green lentil flours*. University of Saskatchewan.

Rabah, R. R. A., Sebastian, A., Vunnam, S., Sultan, S., Tarazi, H., Anbar, H. S., Shehata, M. K., Zaraei, S.-O., Elgendy, S. M., Al Shamma, S. A., et al. (2022). Design, synthesis, and biological evaluation of a new series of pyrazole derivatives: Discovery of potent and selective JNK3 kinase inhibitors. *Bioorganic & Medicinal Chemistry, 69*, 116894.

Rostoll Cangiano, L. (2023). *Developmental adaptations of immune and metabolic function in calves and the influence of the intestinal microbiota in health and disease*. University of Guelph.

Sakurai, K., Uezu, K., Numata, M., Hasegawa, T., Li, C., Kaneko, K., & Shinkai, S. (2005). β-1, 3-Glucan polysaccharides as novel one-dimensional hosts for DNA/RNA, conjugated polymers and nanoparticles. *Chemical Communications, 35*, 4383–4398.

Shah, A., Ul Ashraf, Z., Gani, A., Masoodi, F. A., & Gani, A. (2022). β-Glucan from mushrooms and dates as a wall material for targeted delivery of model bioactive compound: Nutraceutical profiling and bioavailability. *Ultrasonics Sonochemistry, 82*, 105884. https://doi.org/10.1016/j.ultsonch.2021.105884

Sikora, P., Tosh, S. M., Brummer, Y., & Olsson, O. (2013). Identification of high β-glucan oat lines and localization and chemical characterization of their seed kernel β-glucans. *Food Chemistry, 137*(1–4), 83–91.

Song, X. (2014). *Estrogen receptor beta is a negative regulator of mammary cell proliferation*. The University of Vermont and State Agricultural College.

Sorasitthiyanukarn, F. N., Muangnoi, C., Rojsitthisak, P., & Rojsitthisak, P. (2022). Chitosan oligosaccharide/alginate nanoparticles as an effective carrier for astaxanthin with improving stability, in vitro oral bioaccessibility, and bioavailability. *Food Hydrocolloids, 124*, 107246.

Su, Y. (2021). *Synthesis of β-D-glucan based delivery systems loaded with IONPs and oligonucleotides: A mechanistic study of their cancer immunotherapeutic effects*. The Chinese University of Hong Kong (Hong Kong).

Su, Y., Chen, L., Yang, F., & Cheung, P. C. K. (2021). Beta-d-glucan-based drug delivery system and its potential application in targeting tumor associated macrophages. *Carbohydrate Polymers, 253*, 117258.

Su, Y., Li, X., Lam, K. L., & Cheung, P. C. K. (2020). pH-sensitive PEG-coated hyper-branched β-D-glucan derivative as carrier for CpG oligodeoxynucleotide delivery. *Carbohydrate Polymers, 246*, 116621.

Such, G. K., Yan, Y., Johnston, A. P. R., Gunawan, S. T., & Caruso, F. (2015). Interfacing materials science and biology for drug carrier design. *Advanced Materials, 27*(14), 2278–2297.

Sudhakar, M. P., Ravel, M., & Perumal, K. (2021). Pretreatment and process optimization of bioethanol production from spent biomass of Ganoderma lucidum using Saccharomyces cerevisiae. *Fuel, 306*, 121680.

Sumiya, K., Matsunaga, T., Tanaka, M., Mochizuki, S., & Sakurai, K. (2020). Oligo-DNA stoichiometrically binds β-1, 3-glucan with the best fit length. *Biomacromolecules, 21*(12), 4823–4834.

Tan, Y., Chen, L., Li, K., Lou, B., Liu, Y., & Liu, Z. (2022). Yeast as carrier for drug delivery and vaccine construction. *Journal of Controlled Release, 346*, 358–379.

Teo, P. Y., Cheng, W., Hedrick, J. L., & Yang, Y. Y. (2016). Co-delivery of drugs and plasmid DNA for cancer therapy. *Advanced Drug Delivery Reviews, 98*, 41–63.

Tosh, S. M., Wood, P. J., Wang, Q., & Weisz, J. (2004). Structural characteristics and rheological properties of partially hydrolyzed oat β-glucan: The effects of molecular weight and hydrolysis method. *Carbohydrate Polymers, 55*(4), 425–436.

Tsouris, V., Joo, M. K., Kim, S. H., Kwon, I. C., & Won, Y.-Y. (2014). Nano carriers that enable co-delivery of chemotherapy and RNAi agents for treatment of drug-resistant cancers. *Biotechnology Advances, 32*(5), 1037–1050.

Ul Ashraf, Z., Shah, A., Gani, A., Masoodi, F. A., & Noor, N. (2020). Effect of nano-reduction on properties of β-glucan and its use as encapsulating agent for release of α-tocopherol. *Bioactive Carbohydrates and Dietary Fibre, 24*, 100230.

Vannucci, L., Krizan, J., Sima, P., Stakheev, D., Caja, F., Rajsiglova, L., Horak, V., & Saieh, M. (2013). Immunostimulatory properties and antitumor activities of glucans. *International Journal of Oncology, 43*(2), 357–364.

Villacrez, J. L., Carriazo, J. G., & Osorio, C. (2014). Microencapsulation of Andes berry (Rubus glaucus Benth.) aqueous extract by spray drying. *Food and Bioprocess Technology, 7*, 1445–1456.

Wang, P.-C., Zhao, S., Yang, B.-Y., Wang, Q.-H., & Kuang, H.-X. (2016). Anti-diabetic polysaccharides from natural sources: A review. *Carbohydrate Polymers, 148*, 86–97.

Wood, P. J., Siddiqui, I. R., & Paton, D. (1978). *Extraction of high-viscosity gums from oats*.

Xiao, J., Tian, W., Abdullah, Wang, H., Chen, M., Huang, Q., Zhang, M., Lu, M., Song, M., & Cao, Y. (2022). Updated design strategies for oral delivery systems: Maximized bioefficacy of dietary bioactive compounds achieved by inducing proper digestive fate and sensory attributes. *Critical Reviews in Food Science and Nutrition*, 1–20.

Ye, S., Shah, B. R., Li, J., Liang, H., Zhan, F., Geng, F., & Li, B. (2022). A critical review on interplay between dietary fibers and gut microbiota. *Trends in Food Science & Technology, 124*, 237–249.

Zargar, V., Asghari, M., & Dashti, A. (2015). A review on chitin and chitosan polymers: Structure, chemistry, solubility, derivatives, and applications. *ChemBioEng Reviews, 2*(3), 204–226.

Zeng, J.-B., He, Y.-S., Li, S.-L., & Wang, Y.-Z. (2012). Chitin whiskers: An overview. *Biomacromolecules, 13*(1), 1–11.

Zhang, H., Yang, C., Zhou, W., Luan, Q., Li, W., Deng, Q., Dong, X., Tang, H., & Huang, F. (2018). A pH-responsive gel macrosphere based on sodium alginate and cellulose nanofiber for potential intestinal delivery of probiotics. *ACS Sustainable Chemistry & Engineering, 6*(11), 13924–13931.

Zhang, Y., Kong, H., Fang, Y., Nishinari, K., & Phillips, G. O. (2013). Schizophyllan: A review on its structure, properties, bioactivities and recent developments. *Bioactive Carbohydrates and Dietary Fibre, 1*(1), 53–71.

Zhang, Y., Li, Y., Xia, Q., Liu, L., Wu, Z., & Pan, D. (2021). Recent advances of cereal β-glucan on immunity with gut microbiota regulation functions and its intelligent gelling application. *Critical Reviews in Food Science and Nutrition*, 1–17.

# Chapter 8
# Beta Glucans and Nanotechnology

**Sarushi Rastogi, Shikha Pandhi, Sameer Ahmad, and Mudasir Bashir Mir**

## Abbreviations

| | |
|---|---|
| CDIMS | circadian disruption-induced metabolic syndrome |
| DOX | doxorubicin (DOX) |
| HBBG | highland barley |
| NanoGluCur | curcumin-glucan-nanoparticles |
| TPP | tripolyphosphate |

## 8.1 Introduction

Oats, barley, bacteria, yeast, algae, and mushrooms all contain the water-soluble dietary fibre known as β-glucan. Due to its stability and safety for human consumption, β-glucan is a food additive that is approved by the FDA (US Food and Drug Administration) (Ashraf et al., 2020). However, the native form of β-glucan has some drawbacks. These include limited solubility, decreased bioavailability when given orally, and the inability of some cell lines to effectively absorb substances other than nanoparticles. By applying nanotechnology-based approaches—such as

S. Rastogi
Department of Nutrition and Dietetics, School of Allied Health Sciences, MRIIRS, Faridabad, Delhi, India

S. Pandhi
Department of Dairy Science & Food Technology, Banras Hindu University, Varanasi, UP, India

S. Ahmad
Department of Food Technology, School of Interdisciplinary Sciences and Technology, Jamia Hamdard, New Delhi, Delhi, India

M. B. Mir (✉)
Department of Food Science & Technology, Govt. Degree College, Baramulla, Jammu & Kashmir, India



M. Bashir Mir, S. A. Mir (eds.), *Beta-Glucan: Sources, Properties and Applications*, https://doi.org/10.1007/978-3-031-95788-8_8

nanoencapsulation, nanoparticle formulation, and nanosuspensions—scientists have enhanced the stability, solubility, and targeted delivery of beta-glucans. This convergence opens new frontiers in drug delivery, nutraceutical development, and advanced food systems, offering promising applications in healthcare and biotechnology. Nanotechnology is one of the emerging technologies in food processing. The macromolecule lies in the range of 1–100 nm. At the nanoscale, the food materials have a special activity such as a high surface-to-volume ratio enhancing color, physiochemical properties, solubility, toxicity, and diffusivity in food. In food, the nanostructure of the ingredients and the modified structure has application in food additives, delivery of nutrients, and fillers (Singh et al., 2017).

β-glucan is one of the important natural polysaccharides that may be derived from plant and animal sources such as microorganisms. The minerals and vitamins influence the β-glucan yield, and the maximum yield would be beneficial to utilize at the industrial level. The emerging application in the medical field is drug delivery of bioactive compounds. β-glucan is fundamentally a non-starch polysaccharide comprised of a beta bond, its structure varies from sources however, generally found in combinations of different positions soluble in water and may be linear and branched.Therefore, this chapter highlights some of the methods used for production of β-glucan nanoparticles and its application in food industry.

### *8.1.1 Effect of Nano Synthesis on β-glucan*

The skeleton of β glucan, having beta 1,3 glycosidic bonds may sometimes occur with 1,4 glycosidic bonds depending on the sources of beta-glucan such as yeast, bacteria, mushroom algal, etc. It is recognized by the US FDA as a food additive and is considered as generally recognized as a safe compound (Lee et al., 2019). The demand for novel functional foods has increased. Thus nanotechnology in foods has the potential, to fulfil this demand without affecting the nutritional value of food. The nanoparticle of β-glucan is found to be stable even at higher temperature and pressure. In a study, by Atta-Allah et al. (2023) yeast β-glucan nanoparticles were synthesized from β-glucan biopolymers. It was found that the conversion of β-glucan at the nanoscale increases the bioactivity such as antitumor, immunomodulatory, applications in the field of development of medicine. Apart from that, the biocompatible properties of β-glucan could be used for coating and loading the functional ingredients such as phytochemicals and which may be further used for cytotoxicity of pathogens like E.coli (Li & Cheung, 2019). Moreover, they biosynthesized β-glucan from fruit and vegetable waste using K. lactis (355.5 mg/100 mL). The β-glucan nanoparticle was found in the range of 20–90 nm evaluated in UV spectra with average size of 300 nm as measured using dynamic light scattering (DLS).

Sulfation, carboxymethylation, acetylation, hydroxyethylation, hydroxypropylation, methylation, amination, and esterification are the most active reaction sites for structural alteration in β–glucans. These qualities enable them to be built into β-glucan-based nanoparticle systems with improved physiological properties. In order to boost its bioactivity, β-glucan polymer was converted into nanostructures. By solubilizing β-glucan isolated from the *P. aphanidermatum* cell wall in NaOH under ambient circumstances, nanoparticles were created. The regulated addition of tripolyphosphate (TPP), which gives the produced β-glucan nanoparticles structural integrity and stability, transformed the nanoparticles. β-glucan nanoparticles were seen to be round, have a smooth surface, and range in size from 20 to 50 nm (Anusuya & Sathiyabama, 2014). Parthasarathy et al. (2021) displayed the simple formation of polydispersed nanoparticles with a mean size of 20–50 nm from glucan obtained from the plant *Gracilaria corticata*.

Nano-glucan nanoparticles with diameters ranging from 11.5 to 24 nm were produced in one study using supercritical $CO_2$. The nano β-glucan derived from oyster mushrooms were predicted to have a high potential for application at industrial scale as an emulsifier and were greater than other polysaccharides, also suggesting high potential as a flocculating agent (Mahmoud et al., 2015). In a different experiment, beta-glucan was developed on fruit and vegetable wastes and whey and then transformed into nanostructure by *Meyerozyma guilliermondii* and *Kluyveromyces lactis*. N-dimethylformamide, a novel stabiliser, was used. The particles in a TEM picture exhibited smooth surfaces and spherical forms. Sizes of the generated nanoparticles ranged widely from 20 to 90 nm (Atta-Allah et al., 2023). Another research suggested increased water solubility of curcumin by around 180 times where, using the nanoprectipation process, β-glucan derived from Vietnamese mushrooms was used to encapsulate curcumin. The generated curcumin-glucan-nanoparticles (NanoGluCur) are spherical in form and 60–70 nm in diameter. NanoGluCur showed anti-tumor promoting effects by reducing tumour size by 59.93% and tumour density by 40.52% (Le et al., 2016).

To transport genetic material to the target cells, Hwang et al. (2018) created β-glucan nanoparticles. Using a siRNA, the β-glucan molecules self-assembled into nanoparticles with diameters of 300–400 nm.

## 8.2 Applications of β-D-glucan Nanoparticles

β-glucans are a class of polysaccharides that are not digestible and can be derived from various sources, such as cereals, algae, yeast, and bacteria, each with unique structures and properties (Salgado et al., 2017). β-glucan has been recognized for its ability to enhance texture, stabilize, and retain water in food systems, making it a valuable ingredient for specialty foods and offering potential health benefits to consumers. Because it is soluble in water, β-glucan can improve functional properties such as rheology and texture in food systems (Kaur et al., 2019). β-glucans have a wide range of applications (Fig. 8.1) in the food and pharmaceutical industries due to their combination of health-promoting and technological properties (Salgado et al., 2017).

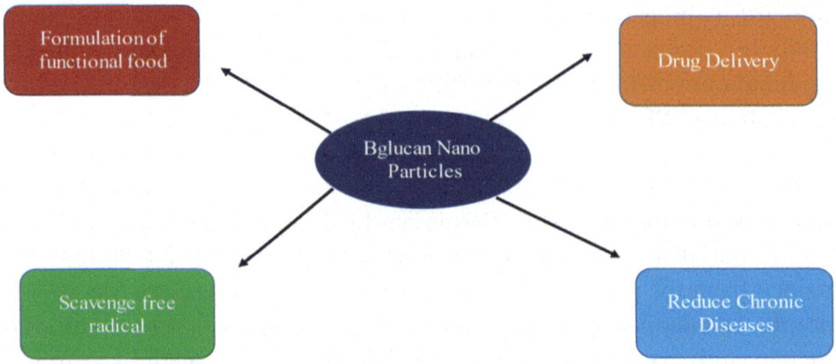


**Fig. 8.1** Application of β-glucan nano particles

### *8.2.1 Beta-glucan Nanoparticles in Drug Delivery*

The potential use of beta-glucan nanoparticles in drug delivery (Fig. 8.1), namely in targeted drug administration and improving cellular absorption, has been thoroughly investigated. They are a desirable choice for the development of innovative treatment techniques due to their unique qualities, which include their capacity to encapsulate different payloads and operate as a particular target for immune cells.

As structural components, 1,3-β-glucans have been utilized to encapsulate insoluble drugs. For example, anti-tumor monoclonal antibodies and yeast-derived beta-glucan have been combined for cancer immunotherapy. It has been demonstrated that oral beta-glucan supplementation amplifies the anti-tumor effects of monoclonal antibodies, emphasizing their potential in cancer treatment. For example, N-dimethylformamide was used as a stabilizer in the synthesis of beta-glucan nanoparticles, producing spherical particles with smooth surfaces and a wide size range. These nanoparticles were less cytotoxic than beta-glucan polymer and showed anticancer efficacy against hepatic cancer (HepG2) cell lines (Atta-Allah et al., 2023).

In another study, low molecular weight beta-glucan was sliced using trifluoroacetic acid (TFA) to create beta-glucan nanoparticles (GluNPs). The nanoparticle formed of the size 250 nm was then employed for immune response strengthening and genetic material delivery. Single strand DNA was inserted into the glucan triple helix structure to verify the effectiveness of GluNPs as a medication or DNA carrier(Hwang et al., 2018).

Besides, targeted drug delivery, Beta-glucan nanoparticles also trigger immune cells, including macrophages, to release immuno-enhancing cytokines, boosting the immune response. This immune activation may be useful in cancer immunotherapy and other therapies that need immune system modification (Huang et al., 2020). However, there are certain constraints when scaling up the production of beta-glucan nanoparticles for drug delivery like maintaining its product characteristics. Preserving the properties of the nanoparticles is essential during the scaling-up procedure, particularly for formulations that are sensitive to pH. For beta-glucan nanoparticles to be used in drug delivery applications, their biocompatibility and safety must be guaranteed on a wider scale (Palol et al., 2023).

### 8.2.2 Beta-glucan Nanoparticles in Delivery of Food Bioactives

Beta-D-glucan nanoparticles as a means of encapsulating food bioactives in nano-delivery systems for agriculture and food has been studied (Sampathkumar et al., 2020). A study demonstrated the use of beta-glucan nanoparticles to encapsulate bioactive substances in order to fortify various food components and create functional meals with improved nutraceutical qualities. Wet milling technology was used to reduce β-glucan into nanoparticles, and catechins were exposed to ultrasonication in order to nanoencapsulate them (Shah et al., 2022). Beta-glucan nanoparticles are superior to other nanoparticles as they maintain the functionality of bioactive substances since they have been demonstrated to improve the bioavailability, stability, and shelf-life of sensitive constituents. Besides, in terms of cost-effectiveness, beta glucan nanoparticles are superior to other nanoparticles. The cost-effective way of manufacturing spherical β-glucan nanoparticles, using nano-precipitation, is its easy one-step process that doesn't require additional heating. Furthermore, the utilization of fungal β-glucans in the creation of biomaterials has advantages including low toxicity, physiological stability, and biodegradability, augmenting their economic viability across a range of uses (Yang & Cheung, 2023).

Encapsulation of bioactive helps in preserving it from harsh environmental conditions in gastrointestinal conditions. β-glucan nanoparticles are ideal delivery vehicles for bioactive substances because they enhance bioavailability and biological activity after digestion while delaying their release in the gastrointestinal system. In a research study, Rutin, a plant-based flavonoid was nano-encapsulated in β-glucan matrix extracted from dates and mushrooms to improve its bioavailability and biological activity after digestion and shield it from the harsh gastrointestinal environment (Shah et al., 2022). β-glucan encapsulation efficiency can be improved by nano-reduction, resulting in food bioactives being released continuously. As shown in a research study where upon reducing to nanosized, its encapsulation efficacy for α-tocopherol increased to 91% for mushroom β-glucan and 89% for date β-glucan. (Ashraf et al., 2020).

### 8.2.3 Beta-glucan Nanoparticles in Reducing Chronic Disease

It has been demonstrated that beta-glucan nanoparticles can protect against chronic illnesses, especially by boosting the immune system. Research discovered that mice were protected against sepsis by trained immunity generated by β-glucan-coupled superparamagnetic iron oxide nanoparticles (BSNPs) (Pan et al., 2022). Besides, the beta-glucan nanoparticles have demonstrated potential to improve the tumor microenvironment for the purpose of cancer immunotherapy. These nanoparticles, which can alter host immune responses by adopting pathogen-associated molecular patterns and can effectively reduce the development of primary tumors and their distant metastases (Zhang et al., 2018).

β-glucans nanoparticles have demonstrated the ability to stimulate immune cell maturation and activation, as well as guide the immune response against tumor cells, suggesting that they might be useful adjuvants in cancer treatment. They can be used to improve tumor regression, immune cell quantity and function, and minimize adverse effects in conjunction with other cancer therapies, such as immunotherapy, chemotherapy, and monoclonal antibodies (Cognigni et al., 2021).

### 8.2.4 Beta-glucan Nanoparticles and Antifungal Activity

Beta-glucan nanoparticles have shown antifungal activity against various fungal species, including *Pythium aphanidermatum* and *Candida albicans*. Their antifungal activity is due to their capacity to damage cell membranes and limit fungal cell development. β-Glucan nanoparticles have also shown positive results in preventing fungal infections against devastating fungus *Pythium aphanidermatum* that affects major crop plants (Anusuya & Sathiyabama, 2014). The antifungal activity of beta-glucan nanoparticles is also being investigated for possible application in agriculture to protect crops against fungal infections. It has been discovered that the nanoparticles may penetrate plant cells and trigger the creation of reactive oxygen species, so activating the plant's defensive mechanism against the fungal disease. Anusuya and Sathiyabama (2015) used β-d-glucan nanoparticles (GNP) in turmeric plants helping in prevention of rhizome rot caused by *Pythium aphanidermatum* in turmeric. The GNP was applied to the turmeric plants via foliar spray, and the results showed improved plant growth, rhizome output, and curcumin content. Furthermore, the treatment resulted in the quick activation of numerous defensive enzymes in the treated plants' leaves and rhizomes, which helped to reduce pathogen colonization.

A recent research study demonstrated that β-glucan-functionalized nanoparticles reduced the inflammatory response in mononuclear phagocytes exposed to *Candida albicans*. By using β1,3-glucan-functionalized polystyrene (β-Glc-PS) nanoparticles, were designed to specifically target the harmful inflammatory effects of fungal infection on the host. The anti-inflammatory cytokines TNF-α, IL-6, IL-1β, and IL-12p40 were observed in the in-vitro culture supernatants of bone marrow-derived dendritic cells and macrophages challenged with *Candida albicans* cells. The study discovered that β-Glc-PS nanoparticles reduced these levels. Furthermore, when bone marrow-derived dendritic cells were cultured with *Candida albicans*, the generation of reactive oxygen species was hindered by β-Glc-PS nanoparticles. The size of the nanoparticles affected this immunomodulatory impact (Lima et al., 2022).

## 8.3 Encapsulation in Nano β-glucan

The β-glucan nanoparticle, compared to the traditional source, has high encapsulation efficiency. Nanoemulsions of β-glucan can resist the change in taste, Odor, color, and sensory properties, even interaction with other food constituents such as vitamins minerals and protein. Moreover, they exhibit compatability with ingredient and efficiently deliver the target compounds and show increased efficacy in the delivery of compounds (Ubbink & Krüger, 2006).

Nanoencapsulation has been done due to the presence of a honeycomb structure of β-glucan which can easily hold bioactive compounds. Chiral active β-glucan has various applications and found to enhance the activity such as work done by Huang et al. (2020). β-glucan extracted from *Saccharomyces cerevisiae was* modified to chiral nanoparticles and found immunological properties which were able to cure the breast cancer and show antitumor efficacy.

## 8.4 Conclusion

The increasing availability of nanoparticles with precise and small dimensions, customised surfaces, enhanced solubility, diversified functionality and shape, and subsequent advances has contributed to the growing interest in polymer nanoparticles. Different methods have been discovered in the development of β-glucan nanoparticles from various edible sources. The nanosized β-glucan was reported to have improved bioactivity, encapsulation efficiency, solubility, antioxidant activity and other functional properties. The potential use of beta-glucan nanoparticles in drug delivery has become a desirable choice for the development of innovative treatment techniques. The beta-glucan nanoparticles have demonstrated potential to improve the tumor microenvironment for the purpose of cancer immunotherapy, and have proven to protect against chronic illnesses, especially by boosting the immune system. The anti-fungal activities of β-glucan-functionalized nanoparticles have also been documented substantially. However, the possible risks of nanoscale materials, as well as the possibility of nanoparticle accumulation and translocation in the body, should be avoided.

## References

Anusuya, S., & Sathiyabama, M. (2014). Preparation of β-d-glucan nanoparticles and its antifungal activity. *International Journal of Biological Macromolecules, 70*, 440–443. https://doi.org/10.1016/j.ijbiomac.2014.07.011

Anusuya, S., & Sathiyabama, M. (2015). Protection of turmeric plants from rhizome rot disease under field conditions by β-d-glucan nanoparticle. *International Journal of Biological Macromolecules, 77*, 9–14. https://doi.org/10.1016/j.ijbiomac.2015.02.053

Ashraf, Z. U., Shah, A., Gani, A., Masoodi, F. A., & Noor, N. (2020). Effect of nano-reduction on properties of β-glucan and its use as encapsulating agent for release of α-tocopherol. *Bioactive Carbohydrates and Dietary Fibre, 24*, 100230. https://doi.org/10.1016/j.bcdf.2020.100230

Atta-Allah, A. A., Ahmed, R. F., Shahin, A. A. M., Hassan, E. A., El-Bialy, H. A.-A., & El-Fouly, M. Z. (2023). Optimizing the synthesis of yeast Beta-glucan via response surface methodology for nanotechnology application. *BMC Microbiology, 23*(1), 110. https://doi.org/10.1186/s12866-023-02845-6

Cognigni, V., Ranallo, N., Tronconi, F., Morgese, F., & Berardi, R. (2021). Potential benefit of β-glucans as adjuvant therapy in immuno-oncology: A review. Exploration of targeted anti-tumor therapy. *Exploration of Targeted Anti-tumor Therapy, 2*(2), 122–138. https://doi.org/10.37349/etat.2021.00036

Huang, J., Wu, C., Tang, S., Zhou, P., Deng, J., Zhang, Z., Wang, Y., & Wang, Z. (2020). Chiral active β-Glucan nanoparticles for synergistic delivery of doxorubicin and immune potentiation. *International Journal of Nanomedicine, 15*, 5083–5095. https://doi.org/10.2147/IJN.S258145

Hwang, J., Lee, K., Gilad, A., & A., & Choi, J. (2018). Synthesis of Beta-glucan nanoparticles for the delivery of single Strand DNA. *Biotechnology and Bioprocess Engineering, 23*(2), 144–149. https://doi.org/10.1007/s12257-018-0003-4

Kaur, R., Sharma, M., Ji, D., Xu, M., & Agyei, D. (2019). Structural features, modification, and functionalities of beta-glucan. *Fibers, 8*(1), 1.

Lee, K., Kwon, Y., Hwang, J., Choi, Y., Kim, K., Koo, H. J., Seo, Y., Jeon, H., & Choi, J. (2019). Synthesis and functionalization of β-Glucan particles for the effective delivery of doxorubicin molecules. *ACS Omega, 4*(1), 668–674. https://doi.org/10.1021/ACSOMEGA.8B02712/ASSET/IMAGES/MEDIUM/AO-2018-02712D_M003.GIF

Li, X., & Cheung, P. C. K. (2019). Application of natural β-glucans as biocompatible functional nanomaterials. *Food Science and Human Wellness, 8*(4), 315–319. https://doi.org/10.1016/J.FSHW.2019.11.005

Le, M. H., Do, H. D., Tran Thi, H. H., Dung, L. V., Nguyen, H. N., Tran Thi, H. N., Nguyen, L. D., Hoang, C. K., Le, H. C., Le Thi, T. H., Trinh, H. T., & Ha, P. T. (2016). The dual effect of curcumin nanoparticles encapsulated by 1-3/1-6 β-glucan from medicinal mushrooms Hericium erinaceus and Ganoderma lucidum. *Advances in Natural Sciences: Nanoscience and Nanotechnology, 7*(4), 045019. https://doi.org/10.1088/2043-6262/7/4/045019

Lima, T., Gunnarsson, S. B., Coelho, E., Evtuguin, D. V., Correia, A., Coimbra, M. A., Cedervall, T., & Vilanova, M. (2022). B-Glucan-functionalized nanoparticles down-modulate the pro-inflammatory response of mononuclear phagocytes challenged with Candida albicans. *Nanomaterials, 12*(14), 2475. https://doi.org/10.3390/nano12142475

Mahmoud, K. F., Amin, A. A., Salama, M. F., & Seliem, E. I. (2015). Encapsulation of nano beta-glucan for preservation of functionality and targeted delivery of bioactive food component. *International Journal of ChemTech Research., 8*, 587–598.

Pan, Y., Li, J., Xia, X., Wang, J., Jiang, Q., Yang, J., Dou, H., Liang, H., Li, K., & Hou, Y. (2022). β-glucan-coupled superparamagnetic iron oxide nanoparticles induce trained immunity to protect mice against sepsis. *Theranostics, 12*(2), 675–688. https://doi.org/10.7150/thno.64874

Parthasarathy, R., Kumar, S. P., Rao, H. Y., & Chelliah, J. (2021). Synthesis of β-Glucan Nanoparticles from Red Algae–Derived β-Glucan for Potential Biomedical Applications. *Applied Biochemistry and Biotechnology, 193*, 3983–3995.

Palol, V. V., Saravanan, S. K., Vuree, S., Chinnadurai, R. K., & Subramanyam, V. (2023). Nanophytosome formulation of β-1,3-glucan and Euglena gracilis extract for drug delivery applications. *MethodsX, 11*, 102480. https://doi.org/10.1016/j.mex.2023.102480

Salgado, M., Santos, F., Rodríguez-Rojo, S., Reis, R. L., Duarte, A. R. C., & Cocero, M. J. (2017). Development of barley and yeast β-glucan aerogels for drug delivery by supercritical fluids. *Journal of CO2 Utilization, 22*, 262–269.

Sampathkumar, K., Tan, K. X., & Loo, S. C. J. (2020). Developing nano-delivery systems for agriculture and food applications with nature-derived polymers. *iScience, 23*(5), 101055. https://doi.org/10.1016/j.isci.2020.101055

Shah, A., Ul Ashraf, Z., Gani, A., Masoodi, F. A., & Gani, A. (2022). B-Glucan from mushrooms and dates as a wall material for targeted delivery of model bioactive compound: Nutraceutical profiling and bioavailability. *Ultrasonics Sonochemistry, 82*, 105884. https://doi.org/10.1016/j.ultsonch.2021.105884

Singh, T., Shukla, S., Kumar, P., Wahla, V., & Bajpai, V. K. (2017). Application of nanotechnology in food science: Perception and overview. *Frontiers in Microbiology, 8*(AUG). https://doi.org/10.3389/FMICB.2017.01501

Ubbink, J., & Krüger, J. (2006). Physical approaches for the delivery of active ingredients in foods. *Trends in Food Science & Technology, 17*(5), 244–254. https://doi.org/10.1016/J.TIFS.2006.01.007

Yang, F., & Cheung, P. C. K. (2023). Fungal β-Glucan-based nanotherapeutics: From fabrication to application. *Journal of Fungi (Basel, Switzerland), 9*(4), 475. https://doi.org/10.3390/jof9040475

Zhang, M., Kim, J. A., & Huang, A. Y.-C. (2018). Optimizing tumor microenvironment for cancer immunotherapy: B-glucan-based nanoparticles. *Frontiers in Immunology, 9*, 341. https://doi.org/10.3389/fimmu.2018.00341

# Chapter 9
# Health Benefits of β –Glucans

**A. Rajalechumi Kamatchi, K. U. Anjali, and Sundaramoorthy Haripriya**

## 9.1 Introduction

The rise of sedentary lifestyles has led to an increase in chronic ailments such as hypertension, diabetes, dyslipidemia, CVD (cardiovascular diseases) and obesity (El Khoury et al., 2012; Maheshwari et al., 2019; Nehmi-Filho et al., 2023). Obesity, physical inactivity, and dietary habits are key risk factors for lifestyle diseases, posing a growing public health challenge worldwide as metabolic diseases significantly impair an individual's quality of life (Yoo & Kim, 2016). South Asians are experiencing a rise in type 2 diabetes and central obesity due to socio-economic advancements and increased consumption of low-priced, highly processed foods. Over one billion people, or two out of every five in Asia, are obese, with over 85% at risk for metabolic illnesses. Insulin resistance mechanisms are linked to metainflammation, a metabolic imbalance linked to obesity and long-term metabolic dysfunction. The metabolic syndrome (MS) is influenced by insulin and leptin signalling pathways and brain-specific nervous system activities (Nehmi-Filho et al., 2023). Asian populations, particularly Chinese and Indians, are at a higher risk for obesity and Type 2 Diabetes mellitus (T2DM), with one-third of Chinese individuals being overweight or obese. Addressing this issue through regular physical activity, weight management, and a healthy diet are essential (Eapen et al., 2009; Therdtatha et al., 2022).

Clinical and animal model (primarily rats and ruminants) studies have shown that consuming adequate amounts of dietary fibre can increase the viscosity of the intestine, faecal bulking, and gut fermentation, leading to the release of short-chain fatty acids (SCFAs), reducing blood glucose and cholesterol levels, and promoting satiety. This has been linked to a reduction in MS cases (Chanmuang et al., 2022; Rahmani et al., 2019). Recent research has also shown that changes in the gut microbiota environment are linked to positive reduction in MS (Chanmuang et al.,

---

A. R. Kamatchi · K. U. Anjali · S. Haripriya (✉)
Department of Food Science and Technology, Pondicherry University, Puducherry, India

 
M. Bashir Mir, S. A. Mir (eds.), *Beta-Glucan: Sources, Properties and Applications*,
https://doi.org/10.1007/978-3-031-95788-8_9

2022; Chen et al., 2022; Shehzad et al., 2023). Beta-glucan (BG) has emerged as a potential solution for treating these disorders, as it reduces cholesterol and triglyceride levels, maintains blood glucose homeostasis, and improves immunity. Beta-glucan is also known for its anti-cancerous and anti-oxidant properties (Chen et al., 2022; Sivieri et al., 2022).

These beneficial dietary fibres are obtained from various sources including cereals, mushrooms, seaweed, and fungi yeast. The physiological activity of beta-glucans is influenced by the type of linkage, whether branched or linear (beta-(1/3), (1/4), and (1/6)), connecting approximately 25,000 glucose monomers (Bobade et al., 2021; Hu et al., 2015; Tiwari & Cummins, 2009). The length and structure of BGs determine their functionality, which varies on considering the source, purification technique, and methods of extraction adopted. Usually, the complex structured BG exhibits enhanced biological activity than the simpler structure (Han et al., 2020; Lante et al., 2023; Mantovani et al., 2008). Insoluble fibre reduces transit time, adds on faecal mass, and improves the removal of bile salts. Soluble dietary fibres, on the other hand, are known for their prolonged satiety, delayed gastric emptying, and lower absorption rate (Bashir & Choi, 2017; Chanmuang et al., 2022; Daou & Zhang, 2012; Würsch & Pi-Sunyer, 1997).

Furthermore, the functionality of beta-glucans differs based on their source and structure. Bacterial BGs exhibit straight linear structure (β-1, 3 glucans) and improved gut health (Laroche & Michaud, 2008). Cereal-based BGs are linear as well, available as β-1, 3 and β-1, 4 links, protecting against cardiovascular disease (CVD), and other health-promoting properties, such as reducing cholesterol and improving overweight individual's insulin sensitivity (Chen & Raymond, 2008; Ho et al., 2016; Lin et al., 2013; Wolever et al., 2021). BG from mushrooms is linked to enhancing immune responses and possesses anti-cancer properties by stimulating immune cells and inhibiting tumour growth (Motta et al., 2021; Savelkoul et al., 2013; Van Steenwijk & Bast, 2021). Algae-derived BG has been shown to offer health advantages such as increasing immune cell activity, lowering inflammation, and improving the anti-bacterial effect against E.coli (Kadam et al., 2015; Kim et al., 2019). Fungal beta glucans specially yeast-derived are linked together by small β-1, 6 and branched β-1, 3 linkages and large β-1,6 and branched β-1,3 linkages respectively, which accelerate and boost the host's immunity, protect upper respiratory tract against infections, improve gut health and reduce gut permeability, and inflammation. (Van Steenwijk & Bast, 2021; Thomas et al., 2022).

The beta-glucans derived from various sources possess beneficial properties as dietary fibres. It has obtained significant consideration for its beneficial health effect in addressing metabolic diseases and improving overall well-being. Other research has shown that BGs activate immune cells to proliferate natural killer cells (NK cells), and macrophages involved in the defence mechanism of the body against infections and diseases (Garcia-Valtanen et al., 2017; Motta et al., 2021; Pelizon et al., 2005). BGs are associated with several health-beneficial characteristics including immunomodulatory, anti-hypertensive, and anti-cancer activities. Still, more studies are to be conducted for a complete understanding of the mechanism behind such potential healthcare benefits, especially for its role in the down regulation of hypertension (Chan et al., 2009; Smith et al., 2008). This chapter would provide wide knowledge to understand these variations and enable targeted utilisation of BGs for specific health benefits, including improvement in gut health, better glucose-insulin response, protection against CVD, ameliorating immune function, improved anti-oxidant, anti-tumour and anti-ageing effects (Fig. 9.1 and Tables 9.1 and 9.2).

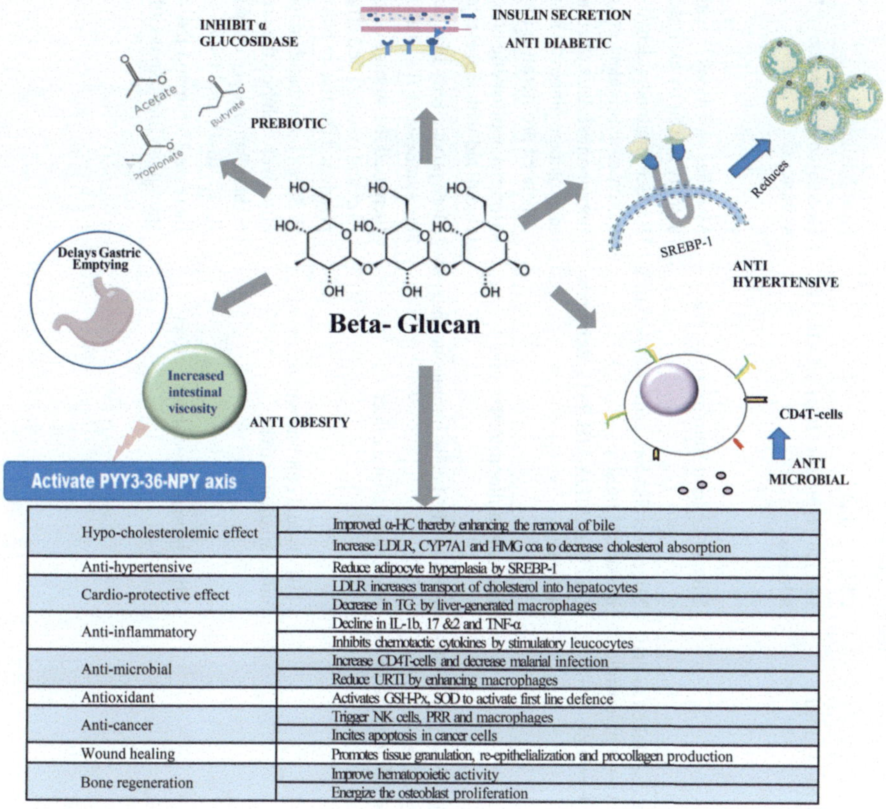


| | |
|---|---|
| Hypo-cholesterolemic effect | Improved α-HC thereby enhancing the removal of bile |
| | Increase LDLR, CYP7A1 and HMG coa to decrease cholesterol absorption |
| Anti-hypertensive | Reduce adipocyte hyperplasia by SREBP-1 |
| Cardio-protective effect | LDLR increases transport of cholesterol into hepatocytes |
| | Decrease in TG by liver-generated macrophages |
| Anti-inflammatory | Decline in IL-1b, 17 &2 and TNF-α |
| | Inhibits chemotactic cytokines by stimulatory leucocytes |
| Anti-microbial | Increase CD4T-cells and decrease malarial infection |
| | Reduce URTI by enhancing macrophages |
| Antioxidant | Activates GSH-Px, SOD to activate first line defence |
| Anti-cancer | Trigger NK cells, PRR and macrophages |
| | Incites apoptosis in cancer cells |
| Wound healing | Promotes tissue granulation, re-epithelialization and procollagen production |
| Bone regeneration | Improve hematopoietic activity |
| | Energize the osteoblast proliferation |

**Fig. 9.1** Beta-glucan and its health benefits. *PRR* Pattern recognition receptors, *PYY* Peptide YY, *GLP1* glucagon-like peptide-1, *PYY3-36-NPY* Peptide YY3-36- neuropeptide Y, *α-HC* alpha 1 microglobulin, *LDLR* low density lipoprotein receptor, *CYP7A1* Cholesterol 7 alpha-hydroxylase, *HMG coa* 3-hydroxy-3-methylglutaryl coenzyme A, *SREBP1* Sterol regulatory element-binding protein-1, *NK* Natural killer, *GSH-Px* Glutathione peroxidase, *SOD* Superoxide dismutase, *CD4T* cluster of differentiation 4 T cells, *URTI* Upper respiratory tract infection, *IL* Interleukins, *TNF* Tumor necrosis factor, *SCFA* Short chain fatty acid, *5-HT* 5-hydroxytryptamine, *TGL* Triglyceride

**Table 9.1** Health benefits of beta-glucan from fungal sources

| Species | Type/composition/structure | Clinical study | Health benefit | Reference |
|---|---|---|---|---|
| *Pleurotus ostreatus* | Beta-(1,3/1,6)-D-glucan | Human-respiratory tract infection in athletes | Decreased occurrence of URTI symptoms | Bergendiova et al. (2011), Majtan (2013), and Nieman et al. (2008) |
| | | Human-cellular immune response | Modifies variations in natural killer cell activity driven by exercise | Bobovčák et al. (2010) |
| *Ganoderma lucidum* | Beta-D-glucose linked to one another by 1 → 3 glycosidic chain with 1 → 6 glycosidic branches | Mice-exercise-inducedoxidative stress in skeletal muscle | Reduces malondialdehyde levels and increases antioxidant enzyme activity | Zhonghui et al. (2014) |
| | | Cell line (MDA-MB-231)-antitumor activity | Immunomodulatory action, including macrophage pinocytic responses and enhanced proliferative impact | Zhao et al. (2010) |
| | | Human-cardioprotective effect | Minimal antidiabetic effect and was able to aid with diabetes'dyslipidemia | Chu et al. (2012) |
| *Cordyceps sinensis* | Every seven (1 → 3)-β-d-Glc*p* backbone residues with two (1 → 6)-β-d-Glc*p* branches | Human PBM cells-immunomodulatory effect | More productivity than cordycepin in the synthesis of IL-10 and IL-2 | Zhou et al. (2008) |
| | | Mice-anti fatigue effect | Increased the storage of glycogen in the liver and muscles and effectively postponed the rise in blood lactic acid | Li and Li (2009) |
| | | THP-1-dual cell line and mice-anti inflammation activity | Reduces intestinal damage and effectively inhibits the production of the main inflammatory cytokines TNF-α, IL-1β, IL-10, and iNOS | Li et al. (2020) |
| | | Mice-immunomodulatory effect | Increase TNF-α and INF-γ levels to strengthen immunomodulatory capacity | Hu et al. (2016) |

| | | | | |
|---|---|---|---|---|
| *Lentinus edodes* | β -(1–6)-glucose branches of five glucose units each on its β-(1–3)-glucose backbone | A549 cells and THP-1 cells-immunomodulatory and pulmonary cytoprotective effects | Decrease oxidative stress, activate macrophages, and lower IL-1β and IL-6 | Murphy et al. (2020a) |
| | | C57BL/6 J male mice-cognition protective effect | Prevents cognitive deficits brought on by a high-fat diet, which might happen through an improvement in the colon-brain axis. | Pan et al. (2021) |
| | | Human-anticancer activity | Beneficial for enhancing quality of life and extending the lifespan of advanced pancreatic patients | Shimizu (2014) |
| | | MCF-7,DU-145, SKOV3 cells-antiproliferative effect | Showed reduced levels of the anti-apoptotic protein Bcl-2 and caspase-3 activation. A promising anti-cancer medication for the efficient treatment of prostate and breast cancer | Titus Sombuor (2022) |
| *Saccharomyces cerevisiae* | A linear backbone of (1,3)-linked D-glucose molecules with (1–6) side chains | A549, MDA-MB-232, and HepG2 -anti cancer effect; resistant *Staphylococcus aureus* (MRSA) and *Pseudomonas aeruginosa*-antimicrobial activity | Offering potential for antibacterial, antitoxic, anticancer, and immune-stimulatory properties | Amer et al. (2021) |
| | | LPS-stimulated RAW264.7 cells-antioxidant effect | Downregulates the generation of reactive oxygen species (ROS) by activating the Nrf2/HO-1 signalling pathway mediated by dendritic-cell-associated C-type lectin 1 (Dectin-1) receptors. | Yu et al. (2021) |
| | | Diabetic male rats-effect on inflammatory or metabolic parameters | Reduces the loss of alveolar bone and enhances the parameters of local and systemic inflammation | Azzi et al. (2021) |

**Table 9.2** Health benefits of beta-glucans from cereal sources

| Species | Type/composition/structure | Clinical study | Health benefit | Reference |
|---|---|---|---|---|
| Barley | Low molecular weight beta-glucan | Male rats-prebiotic effect | Increasing the amounts of Bifidobacterium, bacteroides, and cecal total SCFAs, such as propionate and acetate | Aoe et al. (2021) |
| | (1,3;1,4)-β-glucan | LT97 colon adenoma cells-chemo preventive effect | Increase the shift of molar ratios towards the formation of butyrate, induce the expression of antioxidant and phase II enzymes, and lower the risk of colon cancer | Schlörmann et al. (2020) |
| | (1,3;1,4)-β-glucan | Humans-diabetic patients-hypoglycaemic effect | Increased insulin production and decreased postprandial hyperglycaemia both help to avoid obesity and delay the onset of diabetes-related vascular disease | Fuse et al. (2020) |
| | High molecular weight β-glucan | Humans-cardioprotective effect | The altered microbiome composition is related to a lowering of CVD risk indicators | Wang et al. (2016) |
| | (1,3;1,4)-β-glucan | Metabolic syndrome risk humans- alteration in biomarker levels | Influenced SCFA synthesis, modified the makeup of gut bacteria, and reduced cholesterol levels. | Velikonja et al. (2019) |
| | (1,3;1,4)-β-glucan | C57BL/6 J male mice-hypocholesterolemic effect | SCFAs impact gene expression, resulting in fat build-up, as well as lipid metabolism via gastric absorption and intestinal fermentation | Mio et al. (2020) |
| | (1,3;1,4)-β-glucan | C57BL/6 J mice-effect on gut | Improved insulin sensitivity, decreased appetite, and increased PYY and GLP-1 secretion | Miyamoto et al. (2018) |
| | B-(1 → 4) backbone interrupted by separate β-(1 → 3) linkages | HDFa cells, C57BL/6 wild-type mice | Decreased fibroblast adhesion and dissemination on cell plates, an early reaction in HDF cells that favours migration, and a faster rate of wound healing in mouse skin | Fusté et al. (2019) |

| | | | | |
|---|---|---|---|---|
| Oats | A linear polymer of D-glucose bonded by β-(1 → 4) and β-(1 → 3) glucosidic linkages | Rats-gut microbial changes | Lactobacillus and Bifidobacterium populations have grown and are beneficial for enhancing gut health and function | Shen et al. (2012) |
| | (1 → 3)(1 → 4)-β-glucan | Human- test for satiety | Elevated PYY in circulation and sustained fullness | Beck et al. (n.d.) |
| | (1 → 3)(1 → 4)-β-glucan | Human-anti hypercholesterolemic effect | Decreased weight, improved lipid profile, and lower risk of cardiovascular events | Reyna-Villasmil et al. (2007) |
| | (1 → 3) (1 → 4)-β-D-glucan | Human-weight control | CCK, an appetite suppressor, helps maintain or reduce weight by controlling hunger | Beck et al. (2009) |
| | (1 → 3)(1 → 4)-β-glucan | Human-hypocholesterolemic and prebiotic effect | Increased SCFA synthesis, especially propionate, downregulation of HMG CoA reductase, enhanced cholesterol elimination from cholesterol assimilation, deconjugation of bile acids, and cholesterol binding | Connolly et al. (2016) |
| | (1 → 3)(1 → 4)-β-glucan | Rats-anti microbial and wound healing property | Improved wound healing, showed antibacterial action against *S. aureus* and *M. leutus*, and stopped the proliferation of microorganisms | EL Hosary et al. (2020) |
| | Low molecular oat (1,3) (1,4)-β-d-glucan | Me45, P388/D1, hacat cell line | Robust anti-cancer qualities that are safe and effective to use as an adjuvant in the treatment of skin cancer while posing no hazard to healthy cells | Choromanska et al. (2015) |
| | (1 → 3)(1 → 4)-β-glucan | Male CD-1 mice-immune response | Reduce the elevated risk of URTI brought on by stress from exercise and the rise in macrophage antiviral resistance | Davis et al. (2004) |
| | (1 → 3)(1 → 4)-β-glucan | Human-anti hypertension effect | Significantly lowers both the systolic and diastolic blood pressure and improves blood lipid levels | Keenan et al. (2002) |
| | (1 → 3)(1 → 4)-β-glucan | Human-satiating capacity, appetite and glycaemic control | Reduces hunger and positively modifies postprandial glycaemic levels. No rise in the secretion of plasma GLP-1 | Zaremba et al. (2018) |

## 9.2 Immunomodulation

These beta-glucans come in many different types from multiple sources and provide critical immune-pharmacological functions, including immuno-potentiation, anti-tumour effects, radio-protective, and adjuvant properties. This includes a rise in the host's resistance to infections-fungal, viral, bacterial & protozoal, anti-tumour, anti-carcinogenic, radioprotective, and adjuvant effects (Brown & Gordon, 2003; Petravić-tominac & Panjkota-krbavčić, 2010; Volman et al., 2008; Zeković et al., 2005). BGs have non-specific immunomodulatory actions, stimulating macrophages, NK cells, and antibody synthesis. They are extremely effective in activating the immune response and functions like biological response modifiers (de Oliveira Silva et al., 2017; Thompson et al., 2010). When BG stimulates its receptors, immunological functions improve, such as phagocytosis and proliferation of cytokines (IL-1, IL-6, and GM-CSF). These cytokines stimulate the new formation of white blood cells and also provide power to BG binding receptors seen in most vertebrates (Novak & Vetvicka, 2008; Rodrigues et al., 2020).

The immune-enhancing capability of the fungal mushroom and yeast and algae-derived beta-glucans are well documented both in lab and live models (Chanput et al., 2012; Savelkoul et al., 2013). High MW of BGs (more common in the environment) has been found to associate with the activation of leucocytes directly, boosting their phagocytic and antibacterial capabilities and their cytotoxic behaviour. Intermediate MW of BGs show physiological activity in vivo, despite their less comprehensible cellular effects (Brown & Gordon, 2003; de Oliveira Silva et al., 2017). Macrophages are the primary and best-studied targets for BGs, which activate these cells by increasing their size and number, boosting tumour necrosis factor (TNF) production, and enhancing antigen phagocytosis.

Oral administration of BG solutions derived from fungi decreased alveolar bone loss in rats with gingival inflammation. BGs have long been utilised as an infection-prevention agent (Batbayar et al., 2012; Du et al., 2014; Murphy et al., 2020b). These compounds show protection against infections by bacteria and protozoa in numerous animal models, as well as an enhancement in antimicrobial efficacy towards infections resulting from resistant bacteria. BGs have been studied to protect resistance to *Staphylococcus aureus, Leishmania major, Bacillus anthracis, Trypanosomacruzi, Escherichia coli, L. donovani, Streptococcus suis, Eimeria vermiformis, Candida albicans, Mesocestoidescorti, Plasmodium berghei, and Toxoplasma gondii*, in experimental infectious studies (Chan et al., 2009; de Oliveira Silva et al., 2017; Vetvicka et al., 2019).

The primary public health approach for communities across the world especially those who are more susceptible to infectious illnesses and difficult pathogens is vaccination. To increase the immunogenicity of purified antigens, adjuvants such as glucans, dextrans, lentinans, galactomannans, and glucomannan have been utilized for decades (Van Steenwijk & Bast, 2021; Vetvicka et al., 2020). Because some bacteria, yeast, and seaweed include BGs in their cell walls, researchers are looking at using these polysaccharides as immunostimulatory PRRs (Pattern Recognition

Receptors) for enhancing immunity in the host (Barsanti et al., 2011; de Oliveira Silva et al., 2017; Han et al., 2020).

Beta-glucans can transfer the antigen to cells and trigger an immunological response that is unique to the antigen. Using the spleen tyrosine kinase (Syk)/NF-kB pathway, BGs interact with certain cell receptors (CR3 and Dectin-1), phagosome to produce ROS (reactive oxygen species), internalize pathogens, and produce, pro-inflammatory cytokines (TNF-α, IL-1β, IL-6) (Goodridge et al., 2009; Jang et al., 2009). Influenza A virus is a hazardous infection, and the H5N1 subtype can be directly transferred, and only vaccination can keep such outbreaks under control. Adjuvants are required for vaccine effectiveness. When adjuvants are physiologically efficient, a beneficial interaction occurs involving adjuvant and antigen to induce an innate immune response, eventually leading to adaptive immunity and also used in chemo and radiotherapy for cancer (Barsanti et al., 2011; Motta et al., 2021; Van Steenwijk & Bast, 2021; Vetvicka et al., 2020).

Specially, the BGs from *Saccharomyces cerevisiae* demonstrated effects on mouse in vitro splenic lymphocyte cells, dramatically increasing lymphocyte proliferation and prompting the cells to release IL-2 and INF (Petravić-tominac & Panjkota-krbavčić, 2010; Thomas et al., 2022; Vetvicka, 2011; Vetvicka et al., 2020). The same study looked at adjuvant activity in BALB/c mice inoculated with inactivated H5N1 vaccine *in vivo* and found that fungal BG would significantly enhance the proliferation of lymphocytes, improve CD4+ T cells percentage, bring down the percentage of CD8+ T cells, and escalate the ratio of CD4+/CD8+ (Priyanka et al., 2020; Zhang et al., 2004). *Saccharomyces cerevisiae* BGs are vital bioactive substances for both animal and human health, but their poor solubility has caused certain problems. Its bioactivities might alter and its solubility could be enhanced by sulphate treatment. In an animal model of mice, sulfated glucan was marked to be an efficient adjuvant for H5N1 vaccination. Chitosan-coated PLGA nanoparticles containing 1,3-BG from *Euglena gracilis* functionalized had improved pro-inflammatory response and increased intracellular delivery. The molecular shapes, sizes, and branching of BG molecules affect the adjuvant characteristics differently (Brown & Gordon, 2003; de Oliveira Silva et al., 2017; Vetvicka et al., 2020).

### *9.2.1 Immunomodulatory Mechanism*

The beta-glucan from various sources and the same source-different species provide a variety of BGs with differences in structure, MW and DOP. Through several *in vivo* studies on animals and humans and *in vitro* studies, all types of BGs are found to possess immunomodulatory activity as one of their biological functions. However, due to these differences, the mechanism behind the immunomodulating capacity of BG was not completely elucidated (Petravić-tominac & Panjkota-krbavčić, 2010; Van Steenwijk & Bast, 2021).

Stimulation of a variety of defence-regulating cells and providing immunomodulatory effects is a known fact of BG. It has been discovered that lentinan, a type of BG, increases the cytotoxic activity and synthesis of inflammatory cytokines from immune cells like neutrophils, macrophages and dendritic cells. Additionally, it can promote dendritic cell maturation and improve T-cell performance. However, it is not yet sure whether the T cells are directly activated by lentinan (Chanput et al., 2012; Du et al., 2015; Kim et al., 2011). BGs can elicit immunological responses because they are recognized by PRRs in the inborn immune system. On the surface of inborn immune cells, the PRRs are present to instantly recognize BGs. A class of PRRs, toll-like receptors (TLRs) are found to be present on the cell membrane or in compartments of the endosomal membrane. These TLRs are essential for identifying PAMPs (pathogen-associated molecular patterns) in extracellular space or following endocytosis (Barsanti et al., 2011; Batbayar et al., 2012; Savelkoul et al., 2013).

The potency of BGs to modify immunological responses is related to their interaction with macrophages, which can produce cytokines involved in immune system modulation and regulate phagocytosis (Ciecierska et al., 2019; Goodridge et al., 2009). *In vivo,* investigations have revealed that orally given BGs can stimulate white blood cells responsible for infection defence and tissue healing. Additionally, they promote the synthesis of precursor cells in the bone marrow, which enhances defence against invaders (Motta et al., 2021; Pelizon et al., 2005; Petravić-tominac & Panjkota-krbavčić, 2010; Volman et al., 2008). BGs on oral feeding show interaction with Peyer's patches and intestinal epithelium present in the mucosal layer by the immune cells. Furthermore, they could be picked up by macrophages present in the intestine and reach the spleen, bone marrow, and lymph nodes through transportation (Batbayar et al., 2012; Chan et al., 2009; Priyanka et al., 2020; Volman et al., 2008).

The most common BG targets are the macrophages, which are known to secrete more lysozyme, tumour necrosis factor (TNF), and phagocytose antigens when activated. The NF-kB pathway and TLR2 are involved in the stimulation of macrophages. The TLR2, TLR6, and CD14 receptors work together to regulate the stimulation of macrophages via BGs (Chanput et al., 2012; de Oliveira Silva et al., 2017; Savelkoul et al., 2013; Zeković et al., 2005).

### 9.2.2 Receptors of Beta-Glucan

Investigation on BG recognition revealed several receptors involved in the immune response, including CR3, scavenger receptors, dectin-1, lactosylceramide, and TLRs (Barsanti et al., 2011; Petravić-tominac & Panjkota-krbavčić, 2010; Rodrigues et al., 2020; Volman et al., 2008; Zeković et al., 2005). When these above receptors attach to BGs, they trigger signaling processes and activate immune cells. Numerous studies have strongly established the fact that dectin-1 holds a crucial role in modulating the physiological reaction related to BGs (Brown & Gordon, 2003; Goodridge et al., 2009; Volman et al., 2008). Depending on cell types and the particular receptor activated, the effects of BG identification by various other receptors may also differ. The immune-modulating actions of BGs from fungal and yeast are widely accepted to be mostly dependent on dectin-1, however additional receptors including CR3, TLR2, TLR6, scavenger receptors, or lactosylceramide may also play a role (Geller et al., 2019; Kim et al., 2011; Motta et al., 2021).

Trained immunity via BG conflicts with the belief that innate immunity cells have no immunological memory. Once the immune system is introduced to BG, it can trigger a memory-like response in its cells, enhancing their tendency to fight against infections by activating the PRRs. It is possible through achieving epigenetic reprogramming, increased cytokine production and metabolic alterations. BG used as adjuvant in vaccinations enhances the immune response and promotes long-term immunity with immune memory (Van Steenwijk & Bast, 2021; Thomas et al., 2022).

## 9.3 Anti-diabetic Effect

Diabetes mellitus is a significant public health issue since most of the chronic consequences of the disease make it challenging to carry out everyday activities, and the disease's treatment is significantly costly for society's healthcare system. It has been discovered that about five per cent of the mortality over the world is due to diabetes and also it is expected to affect nearly 366 million people by 2030 and 693 million by 2045 (Andrade et al., 2015; Bozbulut & Sanlier, 2019; Sivieri et al., 2022). Type 1 diabetes mellitus (T1DM) is characterised by no insulin or limited production of insulin due to autoimmune pancreatic beta cell destruction. T2DM is caused by familial and lifestyle variables, including obesity, and has an inheritable polygenic origin. In addition to insulin, hormones involved in glucose metabolism are the incretins and glucagon which are also associated with T2DM. Overproduction of glucagon from alfa cells of the pancreas pumps up the hepatic release of glucose through glycogenolysis and might compromise the peripheral effect of insulin, leading to hyperglycaemia, as this rise in the production of glucose from the liver results in insensitivity to insulin. Advanced lipolysis and escalated levels of free fatty acids (FFA) lead to adipocyte resistance to insulin that further aggravates the failure of beta cells by increasing their resistance to insulin in the liver and muscles (Ferreira et al., 2022; Lante et al., 2023). The GI hormone, incretins- GLP-1 (glucagon-like-peptide 1) and GIP (glucose-dependent insulinotropic polypeptide), on impaired secretion cause disturbance in the glucose utilization. Other factors influencing T2DM include adipocytes' resistance to insulin, which tends to malfunction of beta cells of the pancreas and excess reabsorption of glucose in the kidney by SGLT 2 (sodium/glucose co-transporter 2) contributing to hyperglycaemia (Guo et al., 2023; Lante et al., 2023).

Oral hypoglycaemic medicines and insulin treatment are commonly necessary for the control of diabetes. Uncontrolled diabetes might commonly lead to other health complications like glaucoma, peripheral neuropathy and nephropathy. Low-glycemic-index foods, such as fibre, which includes BG, are essential for diabetes control (Bashir & Choi, 2017; de Oliveira Silva et al., 2017; Sivieri et al., 2022; Yu et al., 2023). The study on BG effects on T1DM is insufficient since this individual depends only on external insulin support and this makes it difficult to evaluate the efficiency of BG. Even though few animal studies delivered positive results for the BG efficiency over T1DM by showing improved glycaemic control, the decline in oxidative stress enhances various systematic mechanisms for developing resistance against several other diabetic complications like CVD, nephropathy, neuropathy and hepatopathy (Bozbulut & Sanlier, 2019; Chen & Seviour, 2007; Guo et al., 2023; Imam et al., 2012).

A huge number of studies including animal and human trials along with *in vitro* studies have proven to reduce the incidence or to manage T2DM with the help of soluble and insoluble BG (Bozbulut & Sanlier, 2019; Chakraborty & Rajeswari, 2022). The gut is involved in insulin secretion management and various mechanisms contributing to T2DM that eventually complement each other since the microbiota can cause changes in the permeability of the intestine, glucose metabolism, inflammatory response, sensitivity towards insulin and energy utilization that are connected to the hyperglycaemic control, reduction in insulin resistance, and lesser cholesterol in plasma (Sivieri et al., 2022; Würsch & Pi-Sunyer, 1997). Generally, having viscous gel-forming capabilities, they play major in controlling glucose and insulin release by reducing the digestive enzymes interaction with food hydrolysates, limiting the transport of glucose to the enterocytes by inhibiting the type 2 glucose transporter (GLUT-2) and SGLT-1 (Bobade et al., 2021; Bozbulut & Sanlier, 2019; Chen et al., 2022). As a result, hypoglycaemic condition achieved via BG with delayed stomach emptying, slower digestion and nutrient absorption, diminished enzymatic activity, hampered chyme transit through the intestine, and limited mucosal absorption (Chakraborty & Devi Rajeswari, 2022; Ciecierska et al., 2019; Mohiuddin, 2022).

Another reason behind BG's hypoglycaemic behaviour is the synthesis of SCFAs from fermentation. These acids trigger peroxisome proliferator receptor-$\gamma$ (PPAR-$\gamma$) and improve the insulin-sensitive type 4 GLUT expression. These GLUT-4 reduces the plasma blood glucose by initiating the transport of glucose to the adipose tissue (Bozbulut & Sanlier, 2019; Chen et al., 2022; Lante et al., 2023; Singh & Bhardwaj, 2023). The impact of BG on the homeostasis of glucose and insulin is possible by the activation of the PI3K/Akt signalling pathway which has a remarkable role in the development of T2DM. Usually, the PI3K/Akt activity is found to be diminished in diabetic cases and studies have proven to increase its activity by administrating BG (Bozbulut & Sanlier, 2019; Chakraborty & Devi Rajeswari, 2022). Furthermore, BG is also involved in appetite and satiety by influencing gut hormone levels like ghrelin and YY. In a human study with 37 diabetic volunteers supplementation with BG was found to show an increase in peptide (PYY) and reduction in GLP-1 thereby improving the insulin levels. Technically, the branched-chain amino acids (BCAA) are considered biomarkers in the diagnosis of diabetes as they often contribute to insulin resistance and hyperglycaemic conditions. Infusion of BG regularly has been reported to decrease the BCAA and also increase insulin resistance (Chen et al., 2022; Guo et al., 2023; Singh & Bhardwaj, 2023).

Diabetes results in dysfunction in the TCA cycle and that leads to hyperglycaemia. Disorders in the TCA cycle caused a reduction in the succinic, citric and hippuric acids and thereby contributed to hyperinsulinemia. Consumption of BG was found to render a hypoglycaemic effect and improve the secretion of insulin through elevation in the expression of succinic and hippuric acids (Guo et al., 2023). Abnormal glucose metabolism in diabetes also affects bile acid metabolism, leading to imbalanced proportions of different bile acids. BG taken as supplements was discovered to regulate bile acid metabolism and improve insulin sensitivity. Additionally, BG affects receptors like SGLT1 and GLUT2, inhibits α-glycosidase enzyme activity, and increases the production of GLP-1, which triggers insulin release and regulates carbohydrate absorption (Bozbulut & Sanlier, 2019; Guo et al., 2023; Lante et al., 2023).

Mostly anti-diabetic and hypoglycaemic effect of BG was found to be from the cereal grains, especially barley, oats and rye. This beta-glucan has evidence to have an effective role in insulin resistance in both diabetic and non-diabetic individuals. The efficiency of beta-glucan to render its protection against diabetes depends on its molecular weight, type of food form, biological interaction, and especially its dosage (Bashir & Choi, 2017; El Khoury et al., 2012; Noronha et al., 2022). Higher molecular weight and highly viscous BGs were discovered to have a hypoglycaemic effect and has been found that viscosity strength is proportional to MW of BG (Noronha et al., 2022; Singh & Bhardwaj, 2023). There are also some studies which conversely state that molecular weight has no remarkable interference in the functional characteristics of BGs as the low molecular weighed BGs resisted small intestine enzymes and produced more SCFAs than the high molecular weighed BGs in a 24 hr. *in vitro* fermentation (Barsanti et al., 2011; Lante et al., 2023).

On coming to the dosage, a rat study (n = 9) was subjected to intravenous intake of 1 mg BG per kg of body weight for a time course of one week and this showed a reduction from 43.30% to 6.70% in the occurrence of diabetes and 84.40% to 26.30% in the insulitis prevalence. Out of nine test rats, eight of them were found to be non-diabetic for about five weeks after quitting the intravenous intake of BG (Najm, 2012). Black yeast-derived BG polycan on supplementation to hamsters with induced hyperlipidemia and liver dysfunction showed anti-diabetic effects without influencing the body weight and diet. Along with the hypoglycaemic effect these polycan also deliver liver repair and control over hyperlipidemia in the hamster (Bashir & Choi, 2017). In another animal study, pigs were chosen and provided with 6% oat BG in their meal, this increased the viscosity of food components in the intestine making it hard for enzymes to reach the nutrients to break and get absorbed thereby increasing the SCFAs synthesis in caecum and colon which results in improved production of insulin and reduction in blood glucose level in pigs (Singh & Bhardwaj, 2023; Suchecka et al., 2017). A test population of non-insulin-dependent diabetic men and women of age 49–57 with an average BMI between 22.6 and 38.9 kg/m$^2$ were analysed for postprandial glycaemic response and insulin response with the cereal breakfast that contains oat BG at various concentrations.

The test population showed an increase in plasma glucose level and post-prandial insulin secretion depending on the dosage of BG ingested. It was found that a 50% reduction in the glycaemic response was achieved on consuming 35 g of breakfast cereal having 5 g of BG derived from oats (Bashir & Choi, 2017; Najm, 2012; Tappy et al., 1996). A human study involving diabetic individuals expressed positive responses for BG from both oats and barley irrespective of the dose and food form but the efficiency altered with the dose, while studies involving healthy human subjects from all age groups were found to express mixed results where some showed an effective reduction in the glycaemic index, postprandial insulin level, HbA1c and few had no significant difference for test food from the control food. The efficiency of BG from oats was significantly more prominent in all human studies than the barley-derived BGs (Bashir & Choi, 2017).

Various studies stated variety of optimum BG dosages in breakfast cereal, cereal bars, muesli, fortified chapatti, pasta, and some drinks have lower postprandial blood glucose, reduced insulinemia and HbA1c, while fettuccini a type of pasta incorporated with BG provided no significant reduction in their glycaemic level on comparing to the control fettuccini (Bashir & Choi, 2017; Bozbulut & Sanlier, 2019; El Khoury et al., 2012). Studies are showing positive response at 0.2 g and also with 9.4 g i.e. 0.1–20% of BG in the food to exhibit a reduction in the glycaemic index from 5% to 50% depending on the dose. The FDA of the USA in 1997 recommended at least 0.75 g or 3 g per portion or per day respectively to experience the above-mentioned health benefits. The EFSA 2011 accepted the BGs's health claim over diabetes and stated that a diet with 4 g of BG in 30 g of carbohydrates might efficiently lower post-prandial glucose levels in blood (Bozbulut & Sanlier, 2019; Guo et al., 2023; Lante et al., 2023; Noronha et al., 2022). The BG's health claims on type 2 diabetes are proven with several animal and human studies while its effect on type 1 diabetes is yet to be discovered.

## 9.4 Effect on Satiety and Obesity

According to the World Health Organisation (WHO), the excessive buildup of fat cells above normal levels is considered to be indicative of being overweight or obese. Similar to diabetes, obesity was also identified as a primary factor contributing to the rising mortality rate. Over the past decades, its occurrence increased and become the fifth major reason for mortality in the world. Its prevalence is seen in all age groups, while older people and females are highly susceptible (Mathews et al., 2021; Shehzad et al., 2023). Obesity is interconnected with other life-threatening diseases like diabetes, CVD, musculoskeletal disorders and specific types of cancers. There are various causes that result in obesity including lifestyle, health, genetic, socioeconomic, behavioral, social, psychological, and environmental factors. Out of which the primary reason is decreased physical activity and increased high-calorie food (low in fibre) consumption (Dubey et al., 2019; Sengül & Ufuk, 2022). Treatment towards obesity usually focuses on increasing physical activity, and decreasing high-calorie food intake and thereby recommends having a balanced nutritious diet by maximizing the intake of fruits, green leafy vegetables, dry fruits and nuts, whole grains and legumes (El Khoury et al., 2012; Maheshwari et al., 2019; Mathews et al., 2021).

Research works on BG, especially from cereals like barley and oats has been found to have an impeccable role in controlling the prevalence of obesity. The effectiveness of the BG on obesity depends on the diet, type and quantity of BG consumed. Dietary intake of BG through cereals like barley and oats has clearly been shown to reduce body mass, and body part measurements including body mass index, neck and waist circumference, skinfold thickness, and waist-hip ratio (Mathews et al., 2021; Sengül & Ufuk, 2022). Certain animal studies express that BG from oats can combat high-fat-diet-influenced obesity by decreasing the fat deposition over the liver and the size of adipocytes (Bai et al., 2019; Sengül & Ufuk, 2022; Xin-Zhong et al., 2015). Dietary fibre, especially the soluble and insoluble forms of BGs has delivered its efficient action in reducing overweight and obesity by reorganizing gut dysbiosis. The BG from yeast, barley and oats are soluble and discovered to enhance the gut microbiota set-up by improving the essential bacteria and reducing the obesity-related bacteria. Latest scientific reports on insoluble dietary fibre also BG render a significant role in altering the gut microbiota composition effectively (Mo et al., 2022; Sivieri et al., 2022). Disregarding solubility, both forms of BG have the potential to alter and organize the gut microbiota, lipid profiles, glucose and insulin metabolism positively. Human studies vouch for BGs ability to decrease body mass, BMI, anthropometric measurements and total calorie intake (Mathews et al., 2021; Rahmani et al., 2019; Shehzad et al., 2023). In conflict with these positive findings, some animal and clinical studies on BG showed either an increase or no change in weight (Ferreira et al., 2022; Rahmani et al., 2019; Xin-Zhong et al., 2015). Despite these conflicts, BGs render a promising avenue for treating obesity and its associated metabolic syndromes.

The possible mechanism for BG to show its impact on weight loss and obesity comes in inter-related. The basic play was by decreasing the energy intake and increasing the feeling of fullness. This happens as the soluble BG increases the intestinal viscosity by creating a viscous gel matrix in the GI path that delays the gastric transit, and reduces the absorption of fat and other nutrients since the interaction between digestive enzymes is restricted. This intestinal condition tends to provide satiety after meals for a long duration and reduce the hunger and habit of overeating (Bai et al., 2019; Bobade et al., 2021; Mathews et al., 2021; Xin-Zhong et al., 2015). The second important criterion is the BG's interference in gut microbiota and its fermentation in the colon. Usually, these non-digested BG on entering the colon alters the microflora by increasing the microbial count of beneficial microbes including species of *Bifidobacterium* and *Lactobacillus*. Further, fermentation of these dietary fibres produces SCFAs like acetate, propionate and butyrate. This SCFA has a direct role in suppressing appetite and obesity by modulating the hormones and receptors associated with obesity (Bai et al., 2019; El Khoury et al., 2012; Mathews et al., 2021; Xin-Zhong et al., 2015; Yu et al., 2023). The third crucial mechanism is triggering the production of anorectic hormones in the gut to suppress hunger. As the SCFAs are produced in the colon, they enhance the gut hormones release including peptide YY (PYY) and GLP-1. Furthermore, oat BG was especially found to activate the gut-hypothalamic axis, in particular the peptide YY3–36-neuropeptide Y (PYY3–36-NPY) axis (Bai et al., 2019; Lin et al., 2013; Mathews et al., 2021; Xin-Zhong et al., 2015). These gut hormones contribute to diminished hunger, low food intake and similarly weight reduction. Some studies suspect the BG to influence about 300 genes that reduce the accumulated fat on activation. While inactivating over 100 genes shows an escalation in fat accumulation (Bai et al., 2019).

The SCFA also influence the GI tract motility, satiety and hunger through multiple pathways. It stimulates the vagal nerves, interferes with intestinal smooth muscle tone and is indirectly involved in the PYY secretion. These SCFAs especially the butyrate activate the FFA-2 (free fatty acid receptor 2) which modifies serotonin release and GI motility. The SCFAs modulate not only PYY and GLP-1 but also other gut hormones like cholecystokinin (CCK) and ghrelin (El Khoury et al., 2012; Lante et al., 2023). BG consumption enhances the anorectic gut hormone and thereby releases GLP-1 and PYY to provide fullness by suppressing the hunger hormone ghrelin in the plasma to decrease food consumption. Along with this, the BG influences the CCK and pancreatic polypeptide which is produced by the F cells of the pancreas to reduce food intake. Another hormone leptin released by adipose tissue also gets involved in the regulation of appetite to mediate energy balance for a longer duration. Based on these collective scientific reports, the BG shows its positive influence on appetite regulation, by modifying the gut hormones through direct and indirect pathways (El Khoury et al., 2012; Mathews et al., 2021; Xin-Zhong et al., 2015).

## 9.5 Cardio-Protective Effect

Cardiovascular diseases (CVD) are one of the major causes of mortality along with diabetes and obesity with an expected death of around 22.2 million by 2030. There are multiple alarming factors related to CVD including high blood pressure, obesity, dyslipidemia, T2DM, genetic and sedentary lifestyle (Sima et al., 2018; Wouk et al., 2021). An increase in obesity, diabetes and other metabolic syndrome elevates the prevalence of CVD and its related health complications. To reduce the mortality by CVDs it is essential to manage, treat and rectify the alarming threats related to the metabolic syndrome including hyperglycemia, oxidative stress, hypertension, and dyslipidemia. Certain lifestyle modifications and dietary interventions remain an important aspect of its prevention and cure.

### 9.5.1 Hypocholesterolemic Effect

A meal with a higher quantity of fibre, in specific like soluble BG is believed to provide cholesterol-lowering effects for reducing CVD (Othman et al., 2011; Queenan et al., 2007; Wolever et al., 2021). Relying on multiple scientific evidence the Food and Drug Administration (FDA) of the USA, the Joint Health Claims Initiatives (JHCI) of the UK and Health Canada have acknowledged the health claims of BG in reducing the cholesterol level which substantially decreases the risk for CVD when at least 3 grams of BG is daily included in the diet (Daou & Zhang, 2012; Othman et al., 2011; Wolever et al., 2021). The insoluble BG directly gets linked to the bile acids to lower the triglycerides and low-density lipoproteins (LDL). The soluble BG increases intestinal viscosity and interferes with cholesterol metabolism by altering the bile acid composition and lipid absorption. These viscous fibres bind to the bile acid, free fatty acids, monoglycerides and cholesterol, inhibiting their uptake in the intestine and further excreted along with the faeces thereby efficiently reducing the circulating lipids. These fibres are found to associate with the production of 7α-hydroxy-4-cholesten-3-one, (α-HC) a metabolite that bulks the removal of bile (Bobade et al., 2021; Ciecierska et al., 2019; Queenan et al., 2007; Sima et al., 2018; Wouk et al., 2021). Apart from this, the cholesterol levels are reduced by influencing the essential colonic bacteria to enhance SCFA production which indirectly involves cholesterol elimination and improved cardiac health (Daou & Zhang, 2012; Queenan et al., 2007; Wolever et al., 2021).

There are multiple direct and indirect pathways in which BGs are involved in treating dyslipidemia, reducing triglycerides and cholesterol levels, and potentially providing cardiac care and preventing CVD (Daou & Zhang, 2012; Sengül & Ufuk, 2022). It is either a cyclic process or the sum of several pathways to exhibit a hypocholesterolemic effect. On consumption of beta-glucan, increase in the LDLR (low-density lipoprotein receptor), CYP7A1 (7-α-hydroxylase), and HMG CoA (3-hydroxy-3-methyl glutaryl-coenzyme A) reductase and decreases the absorption of cholesterol & bile acid and regulates CYP3A4 (Cytochrome P450 3A4). Reduction in the bile acid absorption and promotion of the CYP7A1 leads to an increase in the conversion of the serum cholesterol to bile acid and results in decreased blood cholesterol. LDLR increases the transport of cholesterol into hepatocytes. CYP7A1 escalates the bile acid metabolism while the HMG CoA reductase triggers the cholesterol synthesis (Daou & Zhang, 2012).

A comparative study between BG and a synthetic drug statin against the serum cholesterol showed better results for a statin to bring down TGL (triglyceride), LDL and total cholesterol but caused side effects like intestinal dysfunction, rhabdomyolysis and inflammation, while BG cholesterol-lowering activity was gradual with no side effects (Sima et al., 2018; Wouk et al., 2021). Research findings have established that serum cholesterol levels can be reduced by BG in humans and animal models including rainbow trout, hamsters, mice, rabbits and rats (Othman et al., 2011; Sima et al., 2018; Wood, 2010). BG manages to enhance the GLP-1 secretion to suppress appetite by the L-cells in the intestinal epithelium. It also reduces gluconeogenesis by the hepatic system, lowers the death of beta cells from the pancreas, prevents heart tissue damage, and elevates cardiac debt to improve cardiac function and protect against CVD (Queenan et al., 2007; Yoo & Kim, 2016).

The TGL levels can be decreased by liver-generated macrophages, decreased activity of HMG-CoA reductase and improved CYP7A1 activity in the hepatic system by employing BG (Daou & Zhang, 2012; Othman et al., 2011). The reduction in the adipocyte hyperplasia, body mass and epididymal fat are related to the SREBP-1(sterol regulatory element-binding protein), downward regulation of FAS (fatty acid synthase), and upward regulation of PPAR-α. The AMPK (adenosine monophosphate-activated protein kinase) signalling from the tissues of fat and liver contributed to the upward regulation (Drozdowski et al., 2010; Mishra, 2020; Wouk et al., 2021; Yoo & Kim, 2016). Mushroom-derived BGs improve LDL-receptor mRNA expression in the liver, leading to a decrease in VLDL (very low-density lipoprotein), LDL, TGL, and IDL (intermediate-density lipoprotein), while increasing HDL (high-density lipoprotein) levels to address dyslipidemia. The LDL cholesterol can also be downregulated by a reduction in enterohepatic recirculation of lipids and bile salts (Daou & Zhang, 2012; Queenan et al., 2007). The viscous soluble BG promotes the excretion of cholesterol and other fat substances, thereby increasing cholesterol catabolism, and bile acid production, and reducing lipoprotein secretion, ultimately lowering total lipid levels (Queenan et al., 2007; Wouk et al., 2021).

### 9.5.2 Anti-hypertensive Effect

The occurrence of CVD can also be attenuated by reducing hypertension. In a clinical study, BG inclusion in the diet for about 5% resulted in a reduction of average systemic blood pressure. The BG efficiency towards hypertension is still not well established but this reduction in BP may be related to the viscosity of BG, as it also reduces body weight and cholesterol which sequentially eliminates CVD risk factors. Along with this, the interference of the BG in the reabsorption of sodium in the kidney and the transportation of ions might also cause a decrease in blood pressure (Bobade et al., 2021; Chen et al., 2022; Chen & Raymond, 2008; Chen & Seviour, 2007; Cloetens et al., 2012). Thus BG's positive effect on body mass, total cholesterol, hypertension, and diabetes combined to attenuate the risk of CVD.

Often oxidative stress was claimed to be a core reason for the development of heart diseases. The antioxidant potential of BG from all its sources is proven to provide better protection against the reactive oxygen species (ROS) and deliver efficient scavenging activity against various free radicals including hydrogen peroxide, ABTS, DPPH, hydroxyl ions, and nitric oxide. Mostly the free radicals provoke dysfunction of myocytes and improper functioning of cardiac muscles leading to heart failure and blood pressure abnormalities. Though the BG antioxidant activity is not directly related to CVD, various studies are reporting the BG antioxidant activity to treat and prevent CVD by encountering the ROS (Ciecierska et al., 2019; Queenan et al., 2007; Sengül & Ufuk, 2022; Wouk et al., 2021).

## 9.6 Prebiotic Effect

The number of unhealthy population keep on increasing and rarely could find a healthier individual to satisfy the definition of health as per WHO. This pathetic situation arose as a result of various environmental factors, diet patterns, socioeconomic factors and lifestyle (Maheshwari et al., 2019; Yoo & Kim, 2016). People often suffer from various ailments from gastrointestinal disorders (inflammatory bowel disease, constipation, irritable bowel syndrome-IBS, and diarrhoea), decreased immunity, metabolic disorders (obesity, CVD, diabetes, hypertension, cancer), psychological health disorders (obsessive-compulsive disorder-OCD, depression, anxiety). Upon making necessary changes in lifestyle and dietary patterns gut health can be improved which then works on both physical and mental well-being (Therdtatha et al., 2022; Yoo & Kim, 2016). The importance of the gut is not narrowed down to the digestive system alone but also plays a crucial role in maintaining complete well-being by involving in various aspects of health (Jayachandran et al., 2018; Therdtatha et al., 2022; Yu et al., 2023).

Beta-glucans have earned a significant position as a potential prebiotic in strengthening gut health. Multiple *in vitro* and *in vivo* research findings demonstrated its effect on glucose metabolism, gut composition and its functions (Ciecierska et al., 2019; Jayachandran et al., 2018). After BG supplementation, preclinical and clinical studies delivered strong results indicating its efficiency in reducing diabetes, cholesterol and obesity through modifying the gut microbiota with increment in beneficial bacteria, *Bifidobacteria* and *Lactobacilli* as well as improvement in the *Clostridium perfringens* count. Total bacterial count, enterobacterial and Bacteroides count remained under control while the *enterococci*, *E.coli* and *Clostridium claim* declined in count. *In vivo* studies discovered non-specific fermentation of BG by different *Bifidobacterium species* and *Lactobacilli strains* utilized the BG for their development. Rats given with BG incorporated diets showed an increase in SCFA levels in the caecal, colon, and faeces that provide additional health benefits like hypocholesterolemic effect, immune system stimulation and biofortification (Carlson et al., 2017; Laroche & Michaud, 2008; Suchecka et al., 2017). Understanding the impression of BG prebiotic mechanism towards gut health can provide insights into their potential as dietary interventions for numerous health complications (de Oliveira Silva et al., 2017; Fernandez-Julia et al., 2021; Wouk et al., 2021).

The prebiotic effect of BG was developed after their fermentation by the intestinal bacteria (Daou & Zhang, 2012; Shoukat & Sorrentino, 2021; Wang et al., 2020). The large intestine contains bacteria from various communities that break the non-digestible BG and further ferment leading to a decrease in MW of BG and low MW fragments. These fragments are hydrolyzed by the gut bacteria by using polysaccharide degrading enzymes (Wang et al., 2020). Furthermore, after fermentation, it results in the production of SCFA leading to a consistent decline in pH indicating the metabolic activity of gut microbiota. These SCFA are positively involved in energy production, alteration in lipid metabolism, amino acid metabolism, weight management and managing barrier function of the intestine. These effects contribute to preventing and managing MS (Chen et al., 2022; Hu et al., 2015; Murphy et al., 2020b; Wang et al., 2020).

Among the three primary SCFA, acetate is readily taken up in the peripheral circulation, propionate is removed by the liver and only butyrate is used as an energy substrate by colonocytes. Animal and clinical trials established a specific relationship between SCFA and appetite regulation through various mechanisms. Primarily SCFA contributes to delayed GI motility, limits digestion-nutrient absorption and provides an anorexigenic effect (Mathews et al., 2021; Shehzad et al., 2023). This effect includes the modification in intestinal smooth muscle tone, vagal nerve stimulation in the gut and changes in PYY secretion which all together make an impact on GI motility. These SCFA also regulate GI motility by interacting with the release of 5-hydroxytryptamine (5-HT or serotonin), a potential chemical messenger involved in central and peripheral neural functions by activating the FFA-2 to stimulate vagal nodose neuron to affect colonic transit time (Lin et al., 2013; Shehzad et al., 2023; Singh & Bhardwaj, 2023). Along with PYY, the SCFA can regulate other appetite-related functional hormones including GLP-1, CCK, and ghrelin to provide a feeling of fullness, reduce food intake and also show improved insulin sensitivity and aid weight reduction. Several scientific research data mention the prebiotic effect of BG concerning obesity and gut microbiota. Beneficial change in the gut environment stimulates the immune cell activity, and synthesis of serum amyloid A3 protein (SAA3) that regulates inflammation and adipose tissue to decrease the weight gain and fat deposit in obese individuals that simultaneously lessen central obesity. Gut-produced SCFA also affects leptin secretion, adipogenesis and lipolysis in adipose tissue (Chakraborty & Devi Rajeswari, 2022; Du et al., 2014; El Khoury et al., 2012; Yoo & Kim, 2016).

SCFA may also prevent low-level inflammatory responses and pro-inflammatory adipokines, reduce fasting plasma glucose, and increase the antioxidant effect, especially in type 2 diabetic patients (Shoukat & Sorrentino, 2021; Sivieri et al., 2022; Yoo & Kim, 2016). BG prebiotic effect in animal and human studies shows a hypocholesterolemic effect by reducing the serum TGL, and LDL and increasing HDL levels which thereby reduces dyslipidemia and prevents CVD. These effects are achieved by eliminating the cholesterol and bile in the faeces through the activation of BSH (bile-salt hydrolase) and SCFA production through fermentation (Maheshwari et al., 2019; Therdtatha et al., 2022; Yoo & Kim, 2016). These prebiotic BGs are to be efficiently used by the food formulation industries to develop functional foods that improve overall health aspects.

## 9.7 Other Biological Benefits

The beta-glucans from various sources also exhibit several other health benefits such as antioxidant activities, anti-tumour, anti-microbial, and anti-inflammatory (Chen & Seviour, 2007; El Khoury et al., 2012; Fernandez-Julia et al., 2021;

Goodridge et al., 2009; Pelizon et al., 2005; Sivieri et al., 2022; Van Steenwijk & Bast, 2021). They also work effectively against cancer, help wound healing, and bone formation and protect skin (Du et al., 2014; Liu et al., 2009; Maheshwari et al., 2019; Majtan & Jesenak, 2018).

### *9.7.1 Anti-inflammatory Effect*

The anti-inflammation effect of yeast and oats-derived BG is extensively studied. On oral consumption, though the BG increases inflammation in the mucosal lining of the intestine triggered by DSS (dextran sulphate sodium), its anti-inflammation property aids in recovering the DSS-triggered intestinal inflammation. The mechanism behind this is the capacity of BG to modify inflammatory mediators and enhance the tight protein junctions that influence the permeability of the intestine (Bobade et al., 2021; Lee et al., 2021). Alzheimer's disease and Parkinson's disease can be prevented and treated using BG by activating its anti-inflammatory properties (Mirończuk-Chodakowska et al., 2021). Rat model studies exhibited effective protection against oxidation and decreased enteritis in rats. Various species of mushroom have been shown to inhibit the inflammatory markers and pro-inflammatory, chemotactic cytokines by stimulating the leucocytes. It showed a decline in the IL-1b, 17&2, and TNF-α and also inhibits the secretion of faecal calprotectin and pro-inflammatory cytokines in individuals with inflammatory bowel disease (Du et al., 2015; Priyanka et al., 2020).

### *9.7.2 Anti-infective Effect*

Beta-glucan delivers its defence against infections provided by microbial communities including bacteria, fungi and viruses as it resists the pathogens by enhancing the leucocyte's anti-infective mechanism to promote phagocytosis for elimination of harmful microbes (de Oliveira Silva et al., 2017; Murphy et al., 2020b; Yun et al., 2003). Diverse clinical trials have explored their protection against infection even with high-risk surgery patients, also decreased the use of antibiotics and shortened the need for intensive care unit supervision (Brown & Gordon, 2003; Goodridge et al., 2009; Volman et al., 2008). In cases with a viral infection, the beta-glucan increased the CD4T-cells in AIDS patients, decreased the severity of infections in malaria-affected patients, and also reduced the infection in the upper respiratory tract. It also played a positive role in decreasing the illness and death rate in COVID-19-affected patients (Mirończuk-Chodakowska et al., 2021; Priyanka et al., 2020). BGs are found to provide anti-bacterial against *Mycobacterium tuberculosis* and *Bacillus anthracis*, and anti-fungal protection against *Candida albicans* and *Aspergillus fumigatus* (Chen & Seviour, 2007; Petravić-tominac & Panjkota-krbavčić, 2010; Thompson et al., 2010).

### 9.7.3 Antioxidant Effect

The antioxidant capacity of BG is demonstrated by both *in vivo* and *in vitro* methods and showed its ability to counteract the chain reaction triggered by ROS (Du et al., 2014; Sengül & Ufuk, 2022). They are potential enough to scavenge the free radicals and chelate transition metals. The ability to reduce lipid peroxidation is achieved through inhibiting oxidative stress and activating first-line defence antioxidant enzymes like GSH-Px (glutathione peroxidase), SOD (superoxide dismutase) (Kanagasabapathy et al., 2013; Wu et al., 2011). They were also found to stave off and repair the damage caused by oxidative stress in the kidney and liver using animal models. They also provide optimal anti-ageing effects as they efficiently nullify the ROS and oxidative stress. The studies on polyphenol: BG complexes rendered excellent antioxidant properties and give a novel strategy to enhance the polyphenol's bioavailability and improve their bioactivities (Mirończuk-Chodakowska et al., 2021; Wu et al., 2011).

### 9.7.4 Cell Apoptotic Effect

Beta-glucans have delivered their potency to trigger cell cycle arrest, hinder cell proliferation, and incite apoptosis in diverse cancer cell lines such as colon, liver, lung, lymphoma, melanoma and pancreatic cells (Chen & Seviour, 2007; Priyanka et al., 2020; Zhang et al., 2004). As discussed earlier they act as immunostimulants to trigger and recruit NK cells, PRR, and macrophages as immune effector cells to provide defence against tumours and cancer cells. In vivo, studies on animal models and clinical trials showed their brilliance in stimulating innate and adaptive immunity for inhibiting tumour proliferation and metastasis (de Oliveira Silva et al., 2017; Wani et al., 2021). In studies performed with advanced breast cancer patients promising outcomes are received with improved defence mechanisms by immune cells to reduce tumour growth, especially in cancer patients. Monoclonal antibody therapy and BG synergize together to render enhancement in its therapeutic efficacy in neuroblastoma tumour regression in mice. The immunomodulatory function and anti-tumour properties make BG to be a prominent anti-cancer agent (Chen & Seviour, 2007; Mirończuk-Chodakowska et al., 2021; Petravić-tominac & Panjkota-krbavčić, 2010).

### 9.7.5 Anti-osteoporosis and Bone Regeneration Effect

Several findings reported beneficial effects of BG over osteoporosis to inhibit bone resorption and increase bone regeneration. *In vitro* studies on BG have positively discussed its capacity to improve hematopoietic activity and energize the osteoblast proliferation that boosts bone formation and hinders bone resorption. Additionally, animal studies found to activate bone regeneration, integrate with bone tissues and improve bone mineralization. Clinical trials have also found a prominent proliferation of markers involved in bone metabolism and bone density (Bashir & Choi, 2017; Jung et al., 2016).

### *9.7.6 Wound Healing Property*

Beta-glucans also express wound healing properties by enhancing macrophage's movement towards the injury site, promoting tissue granulation, re-epithelialization, and production of procollagen thereby improving the deposit of collagen. In vitro studies on BG exhibited improvement in wound closure, cell migration, and fibroblast keratinocyte proliferation. Animal and human studies confirmed the positive effects of beta-glucan in treating chronic wounds and burned injuries (Du et al., 2014; Majtan & Jesenak, 2018; Petravić-tominac & Panjkota-krbavčić, 2010).

### *9.7.7 Anti-aging Property*

Beta-glucan shows protection towards the skin by reducing the depth and height of wrinkles, and roughness, and promoting the synthesis of collagen in the skin's dermal layer (Redmond, 2005). It is possible through its antioxidant properties of BG to inhibit the cross-linked protein formation which causes skin ageing and also protects from UV-induced damage by nullifying the ROS. It also enhances the moisturizing effect by improving hydration and increasing the water level in the outermost layer of the skin-stratum corneum. Due to its evidential skin benefits, cosmetic and pharmaceutical industries are using it as an effective component to slow down ageing, and promote skin health (Du et al., 2014; Vetvicka et al., 2019).

## 9.8 Conclusion

Beta-glucan has many health advantages and has the potential for use in the food and pharmaceutical sectors. It has garnered considerable interest in the field of study because of its immunomodulatory, antimicrobial, antitumor, anticancer, and antioxidant activities. They have nailed their efficiency in the therapeutic and cosmetic applications, as well as in the control and management of metabolic syndrome (MS) and its related threats. Studies revealed their beneficial effects depend on certain criteria such as dosage, source, and MW and carrier food. Efforts to increase daily consumption of dietary fibre are crucial and research has been carried out to incorporate BG in daily food products. Incorporation of BG-rich source food components in daily diet would find its path to ameliorate health by mitigating the risk factors of metabolic syndrome.

## References


Amer, E. M., Saber, S. H., Markeb, A. A., Elkhawaga, A. A., Mekhemer, I. M. A., Zohri, A. N. A., Abujamel, T. S., Harakeh, S., & Abd-Allah, E. A. (2021). Enhancement of β-glucan biological activity using a modified acid-base extraction method from saccharomyces cerevisiae. *Molecules, 26*(8), 1–17. https://doi.org/10.3390/molecules26082113

Andrade, E. F., Lobato, R. V., De Araújo, T. V., Zangerônimo, M. G., De Sousa, R. V., & Pereira, L. J. (2015). Efecto De Los Beta-Glucanos En El Control De Los Niveles De Glucosa En Pacientes Diabéticos: Revisión Sistemática. *Nutricion Hospitalaria, 31*(1), 170–177. https://doi.org/10.3305/nh.2015.31.1.7597

Aoe, S., Mio, K., Yamanaka, C., & Kuge, T. (2021). Low molecular weight barley β-glucan affects glucose and lipid metabolism by prebiotic effects. *Nutrients, 13*(1), 1–13. https://doi.org/10.3390/nu13010130

Azzi, D. V., de Jesus Pereira, A. N., de Oliveira Silva, V., de Carvalho Foureaux, R., Lima, A. R. V., Barducci, R. S., Albuquerque, A. S., Reis, G. L., de Oliveira, R. R., Andrade, E. F., Zangeronimo, M. G., Chalfun-Júnior, A., & Pereira, L. J. (2021). Dose-response effect of prebiotic ingestion (β-glucans isolated from Saccharomyces cerevisiae) in diabetic rats with periodontal disease. *Diabetology and Metabolic Syndrome, 13*(1), 1–11. https://doi.org/10.1186/s13098-021-00729-1

Bai, J., Ren, Y., Li, Y., Fan, M., Qian, H., Wang, L., Wu, G., Zhang, H., Qi, X., Xu, M., & Rao, Z. (2019). Physiological functionalities and mechanisms of β-glucans. *Trends in Food Science and Technology, 88*(February), 57–66. https://doi.org/10.1016/j.tifs.2019.03.023

Barsanti, L., Passarelli, V., Evangelista, V., Frassanito, A. M., & Gualtieri, P. (2011). Chemistry, physico-chemistry and applications linked to biological activities of β-glucans. *Natural Product Reports, 28*(3), 457–466. https://doi.org/10.1039/c0np00018c

Bashir, K. M. I., & Choi, J. S. (2017). Clinical and physiological perspectives of β-glucans: The past, present, and future. *International Journal of Molecular Sciences, 18*(9). https://doi.org/10.3390/ijms18091906

Batbayar, S., Lee, D. H., & Kim, H. W. (2012). Immunomodulation of fungal β-glucan in host defense signaling by dectin-1. *Biomolecules and Therapeutics, 20*(5), 433–445. https://doi.org/10.4062/biomolther.2012.20.5.433

Beck, E. J., Tosh, S. M., Batterham, M. J., & Tapsell, L. C. (2009). Oat β-glucan increases postprandial cholecystokinin levels, decreases insulin response and extends subjective satiety in overweight subjects. *Molecular Nutrition Food Research, 53*, 1343–1351. https://doi.org/10.1002/mnfr.200800343

Beck, E. J., Tapsell, L. C., Batterham, M. J., Tosh, S. M., & Huang, X.-F. (n.d.). *Increases in peptide Y-Y levels following oat β-glucan ingestion are dose-dependent in overweight adults – ScienceDirect.* https://www.sciencedirect.com/science/article/pii/S0271531709001699?via%3Dihub

Bergendiova, K., Tibenska, E., & Majtan, J. (2011). Pleuran (β-glucan from Pleurotus ostreatus) supplementation, cellular immune response and respiratory tract infections in athletes. *European Journal of Applied Physiology, 111*(9), 2033–2040. https://doi.org/10.1007/s00421-011-1837-z

Bobade, H., Gupta, A., & Sharma, S. (2021). Beta-glucan. *Nutraceuticals and Health Care, January*, 343–358. https://doi.org/10.1016/B978-0-323-89779-2.00013-2

Bobovčák, M., Kuniaková, R., Gabriž, J., & Majtán, J. (2010). Effect of pleuran (β-glucan from pleurotus ostreatus) supplementation on cellular immune response after intensive exercise in elite athletes. *Applied Physiology, Nutrition and Metabolism, 35*(6), 755–762. https://doi.org/10.1139/H10-070

Bozbulut, R., & Sanlier, N. (2019). Promising effects of β-glucans on glyceamic control in diabetes. *Trends in Food Science and Technology, 83*(October 2018), 159–166. https://doi.org/10.1016/j.tifs.2018.11.018

Brown, G. D., & Gordon, S. (2003). Fungal β-glucans and mammalian immunity. *Immunity, 19*(3), 311–315. https://doi.org/10.1016/S1074-7613(03)00233-4

Carlson, J. L., Erickson, J. M., Hess, J. M., Gould, T. J., & Slavin, J. L. (2017). Prebiotic dietary fiber and gut health: Comparing the in vitro fermentations of beta-glucan, inulin and xylooligosaccharide. *Nutrients, 9*(12). https://doi.org/10.3390/nu9121361

Chakraborty, S., & Devi Rajeswari, V. (2022). Biomedical aspects of beta-glucan on glucose metabolism and its role on primary gene PIK3R1. *Journal of Functional Foods, 99*(August), 0–11. https://doi.org/10.1016/j.jff.2022.105296

Chakraborty, S., & Rajeswari, V. D. (2022). *Disease and Insulin resistance-A Perspective Review Therapeutic role of β -glucan on the Association between Polycystic Ovarian Disease and Insulin resistance-A Perspective Review. October*. https://doi.org/10.31838/ijpr/2022.14.04.001

Chan, G. C. F., Chan, W. K., & Sze, D. M. Y. (2009). The effects of beta-glucan on human immune and cancer cells. *Journal of Hematology & Oncology, 2*, 25. https://doi.org/10.1186/1756-8722-2-25

Chanmuang, S., Nguyen, Q. A., & Kim, H. J. (2022). Current research on the effects of non-digestible carbohydrates on metabolic disease. *Applied Sciences (Switzerland), 12*(8). https://doi.org/10.3390/app12083768

Chanput, W., Reitsma, M., Kleinjans, L., Mes, J. J., Savelkoul, H. F. J., & Wichers, H. J. (2012). β-Glucans are involved in immune-modulation of THP-1 macrophages. *Molecular Nutrition and Food Research, 56*(5), 822–833. https://doi.org/10.1002/mnfr.201100715

Chen, J., & Raymond, K. (2008). Beta-glucans in the treatment of diabetes and associated cardiovascular risks. *Vascular Health and Risk Management, 4*(6), 1265–1272. https://doi.org/10.2147/vhrm.s3803

Chen, J., & Seviour, R. (2007). Medicinal importance of fungal β-(1→3), (1→6)-glucans. *Mycological Research, 111*(6), 635–652. https://doi.org/10.1016/j.mycres.2007.02.011

Chen, C., Huang, X., Wang, H., Geng, F., & Nie, S. (2022). Effect of β-glucan on metabolic diseases: A review from the gut microbiota perspective. *Current Opinion in Food Science, 47*(Cvd), 100907. https://doi.org/10.1016/j.cofs.2022.100907

Choromanska, A., Kulbacka, J., Rembialkowska, N., Pilat, J., Oledzki, R., Harasym, J., & Saczko, J. (2015). Anticancer properties of low molecular weight oat beta-glucan – An in vitro study. *International Journal of Biological Macromolecules, 80*, 23–28. https://doi.org/10.1016/j.ijbiomac.2015.05.035

Chu, T. T. W., Benzie, I. F. F., Lam, C. W. K., Fok, B. S. P., Lee, K. K. C., & Tomlinson, B. (2012). Study of potential cardioprotective effects of Ganoderma lucidum (Lingzhi): results of a controlled human intervention trial. *The British Journal of Nutrition, 107*(7), 1017–1027. https://doi.org/10.1017/S0007114511003795

Ciecierska, A., Drywień, M. E., Hamulka, J., & Sadkowski, T. (2019). Nutraceutical functions of beta-glucans in human nutrition. *Roczniki Panstwowego Zakladu Higieny, 70*(4), 315–324. https://doi.org/10.32394/rpzh.2019.0082

Cloetens, L., Ulmius, M., Johansson-Persson, A., Åkesson, B., & Önning, G. (2012). Role of dietary beta-glucans in the prevention of the metabolic syndrome. *Nutrition Reviews, 70*(8), 444–458. https://doi.org/10.1111/j.1753-4887.2012.00494.x

Connolly, M. L., Tzounis, X., Tuohy, K. M., & Lovegrove, J. A. (2016). Hypocholesterolemic and prebiotic effects of a whole-grain oat-based granola breakfast cereal in a cardio-metabolic "at risk" population. *Frontiers in Microbiology, 7*(NOV), 1–9. https://doi.org/10.3389/fmicb.2016.01675

Daou, C., & Zhang, H. (2012). Oat beta-glucan: Its role in health promotion and prevention of diseases. *Comprehensive Reviews in Food Science and Food Safety, 11*(4), 355–365. https://doi.org/10.1111/j.1541-4337.2012.00189.x

Davis, J. M., Murphy, E. A., Brown, A. S., Carmichael, M. D., Ghaffar, A., & Mayer, E. P. (2004). Effects of oat β-glucan on innate immunity and infection after exercise stress. *Medicine and Science in Sports and Exercise, 36*(8), 1321–1327. https://doi.org/10.1249/01.MSS.0000135790.68893.6D

de Oliveira Silva, V., de Moura, N. O., de Oliveira, L. J. R., Peconick, A. P., & José Pereira, L. (2017). Promissing effects of beta-glucans on metabolism and on the immune responses: Review article. *American Journal of Immunology, 13*(1), 62–72. https://doi.org/10.3844/ajisp.2017.62.72

Drozdowski, L. A., Reimer, R. A., Temelli, F., Bell, R. C., Vasanthan, T., & Thomson, A. B. R. (2010). β-Glucan extracts inhibit the in vitro intestinal uptake of long-chain fatty acids and cholesterol and down-regulate genes involved in lipogenesis and lipid transport in rats. *Journal of Nutritional Biochemistry, 21*(8), 695–701. https://doi.org/10.1016/j.jnutbio.2009.04.003

Du, B., Bian, Z., & Xu, B. (2014). Skin health promotion effects of natural Beta-Glucan derived from cereals and microorganisms: A review. *Phytotherapy Research, 28*(2), 159–166. https://doi.org/10.1002/ptr.4963

Du, B., Lin, C., Bian, Z., & Xu, B. (2015). An insight into anti-inflammatory effects of fungal beta-glucans. *Trends in Food Science and Technology, 41*(1), 49–59. https://doi.org/10.1016/j.tifs.2014.09.002

Dubey, S. K., Chaturvedi, V. K., Mishra, D., Bajpeyee, A., Tiwari, A., & Singh, M. P. (2019). Role of edible mushroom as a potent therapeutics for the diabetes and obesity. *3 Biotech, 9*(12), 1–12. https://doi.org/10.1007/s13205-019-1982-3

Eapen, D., Kalra, G. L., Merchant, N., Arora, A., & Khan, B. V. (2009). Metabolic syndrome and cardiovascular disease in South Asians. *Vascular Health and Risk Management, 5*, 731–743. https://doi.org/10.2147/vhrm.s5172

EL Hosary, R., El-Mancy, S. M. S., El Deeb, K. S., Eid, H. H., EL Tantawy, M. E., Shams, M. M., Samir, R., Assar, N. H., & Sleem, A. A. (2020). Efficient wound healing composite hydrogel using Egyptian Avena sativa L. polysaccharide containing β-glucan. *International Journal of Biological Macromolecules, 149*, 1331–1338. https://doi.org/10.1016/j.ijbiomac.2019.11.046

El Khoury, D., Cuda, C., Luhovyy, B. L., & Anderson, G. H. (2012). Beta glucan: Health benefits in obesity and metabolic syndrome. *Journal of Nutrition and Metabolism, 2012*. https://doi.org/10.1155/2012/851362

Fernandez-Julia, P. J., Munoz-Munoz, J., & van Sinderen, D. (2021). A comprehensive review on the impact of β-glucan metabolism by Bacteroides and Bifidobacterium species as members of the gut microbiota. *International Journal of Biological Macromolecules, 181*, 877–889. https://doi.org/10.1016/j.ijbiomac.2021.04.069

Ferreira, C. S., Vendramini, T. H. A., Amaral, A. R., Rentas, M. F., Ernandes, M. C., da Silva, F. L., Oba, P. M., de Oliveira Roberti Filho, F., & Brunetto, M. A. (2022). Metabolic variables of obese dogs with insulin resistance supplemented with yeast beta-glucan. *BMC Veterinary Research, 18*(1), 1–10. https://doi.org/10.1186/s12917-021-03106-2

Fuse, Y., Higa, M., Miyashita, N., Fujitani, A., Yamashita, K., Ichijo, T., Aoe, S., & Hirose, T. (2020). Effect of High β-glucan barley on postprandial blood glucose and insulin levels in type 2 diabetic patients. *Clinical Nutrition Research, 9*(1), 43. https://doi.org/10.7762/cnr.2020.9.1.43

Fusté, N. P., Guasch, M., Guillen, P., Anerillas, C., Cemeli, T., Pedraza, N., Ferrezuelo, F., Encinas, M., Moralejo, M., & Garí, E. (2019). Barley β-glucan accelerates wound healing by favoring migration versus proliferation of human dermal fibroblasts. *Carbohydrate Polymers, 210*(October 2018), 389–398. https://doi.org/10.1016/j.carbpol.2019.01.090

Garcia-Valtanen, P., Guzman-Genuino, R. M., Williams, D. L., Hayball, J. D., & Diener, K. R. (2017). Evaluation of trained immunity by β-1, 3 (d)-glucan on murine monocytes in vitro and duration of response in vivo. *Immunology and Cell Biology, 95*(7), 601–610. https://doi.org/10.1038/icb.2017.13

Geller, A., Shrestha, R., & Yan, J. (2019). Yeast-derived β-glucan in cancer: Novel uses of a traditional therapeutic. *International Journal of Molecular Sciences, 20*(15), 1–20. https://doi.org/10.3390/ijms20153618

Goodridge, H. S., Wolf, A. J., & Underhill, D. M. (2009). β-glucan recognition by the innate immune system. *Immunol Rev, 230*, 38–50.

Guo, H., Wu, H., Kong, X., Zhang, N., Li, H., Dong, X., & Li, Z. (2023). Oat β-glucan ameliorates diabetes in high fat diet and streptozotocin-induced mice by regulating metabolites. *Journal of Nutritional Biochemistry, 113*. https://doi.org/10.1016/j.jnutbio.2022.109251

Han, B., Baruah, K., Cox, E., Vanrompay, D., & Bossier, P. (2020). Structure-functional activity relationship of β-glucans from the perspective of immunomodulation: A mini-review. *Frontiers in Immunology, 11*(April), 1–8. https://doi.org/10.3389/fimmu.2020.00658

Ho, H. V. T., Sievenpiper, J. L., Zurbau, A., Blanco Mejia, S., Jovanovski, E., Au-Yeung, F., Jenkins, A. L., & Vuksan, V. (2016). The effect of oat β-glucan on LDL-cholesterol, non-HDL-cholesterol and apoB for CVD risk reduction: A systematic review and meta-analysis of randomised-controlled trials. *British Journal of Nutrition, 116*(8), 1369–1382. https://doi.org/10.1017/S000711451600341X

Hu, X., Zhao, J., Zhao, Q., & Zheng, J. (2015). Structure and characteristic of β-glucan in cereal: A review. *Journal of Food Processing and Preservation, 39*(6), 3145–3153. https://doi.org/10.1111/jfpp.12384

Hu, T., Jiang, C., Huang, Q., & Sun, F. (2016). A comb-like branched β-d-glucan produced by a Cordyceps sinensis fungus and its protective effect against cyclophosphamide-induced immunosuppression in mice. *Carbohydrate Polymers, 142*, 259–267. https://doi.org/10.1016/j.carbpol.2016.01.036

Imam, M. U., Musa, S. N. A., Azmi, N. H., & Ismail, M. (2012). Effects of white rice, brown rice and germinated brown rice on antioxidant status of type 2 diabetic rats. *International Journal of Molecular Sciences, 13*(10), 12952–12969. https://doi.org/10.3390/ijms131012952

Jang, S. A., Park, S., Lim, J. D., Kang, S. C., Yang, K. H., Pyo, S., & Sohn, E. H. (2009). The comparative immunomodulatory effects of β-glucans from yeast, bacteria, and mushroom on the function of macrophages. *Journal of Food Science and Nutrition, 14*(2), 102–108. https://doi.org/10.3746/jfn.2009.14.2.102

Jayachandran, M., Chen, J., Chung, S. S. M., & Xu, B. (2018). A critical review on the impacts of β-glucans on gut microbiota and human health. *Journal of Nutritional Biochemistry, 61*, 101–110. https://doi.org/10.1016/j.jnutbio.2018.06.010

Jung, M. Y., Kim, J. W., Kim, K. Y., Choi, S. H., & Ku, S. K. (2016). Polycan, a β-glucan from Aureobasidium pullulans SM-2001, mitigates ovariectomy-induced osteoporosis in rats. *Experimental and Therapeutic Medicine, 12*(3), 1251–1262. https://doi.org/10.3892/etm.2016.3485

Kadam, S. U., Tiwari, B. K., & O'Donnell, C. P. (2015). Extraction, structure and biofunctional activities of laminarin from brown algae. *International Journal of Food Science and Technology, 50*(1), 24–31. https://doi.org/10.1111/ijfs.12692

Kanagasabapathy, G., Malek, S. N. A., Mahmood, A. A., Chua, K. H., Vikineswary, S., & Kuppusamy, U. R. (2013). Beta-glucan-rich extract from pleurotus sajor-caju (Fr.) singer prevents obesity and oxidative stress in C57BL/6J mice fed on a high-fat diet. *Evidence-Based Complementary and Alternative Medicine*. https://doi.org/10.1155/2013/185259

Keenan, J. M., Pins, J. J., Frazel, C., Moran, A., & Turnquist, L. (2002). Oat ingestion reduces systolic and diastolic blood pressure in patients with mild or borderline hypertension: A pilot trial. *The Journal of Family Practice, 51*(4), 369.

Kim, H. S., Hong, J. T., Kim, Y., & Han, S.-B. (2011). Stimulatory effect of β-glucans on immune cells. *Immune Network, 11*(4), 191. https://doi.org/10.4110/in.2011.11.4.191

Kim, K., Ehrlich, A., Perng, V., Chase, J. A., Raybould, H., Li, X., Atwill, E. R., Whelan, R., Sokale, A., & Liu, Y. (2019). Algae-derived β-glucan enhanced gut health and immune responses of weaned pigs experimentally infected with a pathogenic E. coli. *Animal Feed Science and Technology, 248*(August 2018), 114–125. https://doi.org/10.1016/j.anifeedsci.2018.12.004

Lante, A., Canazza, E., & Tessari, P. (2023). Beta-glucans of cereals: Functional and technological properties. *Nutrients, 15*(9). https://doi.org/10.3390/nu15092124

Laroche, C., & Michaud, P. (2008). New developments and prospective applications for beta (1,3) glucans. *Recent Patents on Biotechnology, 1*(1), 59–73. https://doi.org/10.2174/187220807779813938

Lee, A., Ko, J., Ahn, S., Kim, H. J., Min, S., & Kim, E. (2021). Potential effects of pigmented rice on immunity: A review focusing on anthocyanins. *Journal of Food and Nutrition Research, 9*(1), 26–31. https://doi.org/10.12691/jfnr-9-1-4

Li, T., & Li, W. (2009). Impact of polysaccharides from Cordyceps on anti-fatigue in mice. *Scientific Research and Essays, 4*(7), 705–709.

Li, L. Q., Song, A. X., Yin, J. Y., Siu, K. C., Wong, W. T., & Wu, J. Y. (2020). Anti-inflammation activity of exopolysaccharides produced by a medicinal fungus Cordyceps sinensis Cs-HK1 in cell and animal models. *International Journal of Biological Macromolecules, 149*, 1042–1050. https://doi.org/10.1016/j.ijbiomac.2020.02.022

Lin, N., Li, Y., Tang, L., Shi, J., & Chen, Y. (2013). In vivo effect of oat cereal β-glucan on metabolic indexes and satiety-related hormones in diet-induced obesity C57-Bl mice. *Molecular Nutrition and Food Research, 57*(7), 1291–1294. https://doi.org/10.1002/mnfr.201200695

Liu, J., Gunn, L., Hansen, R., & Yan, J. (2009). Combined yeast-derived β-glucan with anti-tumor monoclonal antibody for cancer immunotherapy. *Experimental and Molecular Pathology, 86*(3), 208–214. https://doi.org/10.1016/j.yexmp.2009.01.006

Maheshwari, G., Sowrirajan, S., & Joseph, B. (2019). β-Glucan, a dietary fiber in effective prevention of lifestyle diseases – An insight. *Bioactive Carbohydrates and Dietary Fibre, 19*(July 2018). https://doi.org/10.1016/j.bcdf.2019.100187

Majtan, J. (2013). Pleuran (β-glucan from pleurotus ostreatus): An effective nutritional supplement against upper respiratory tract infections? *Medicine and Sport Science, 59*, 57–61. https://doi.org/10.1159/000341967

Majtan, J., & Jesenak, M. (2018). β-glucans: Multi-functional modulator of wound healing. *Molecules, 23*(4), 1–15. https://doi.org/10.3390/molecules23040806

Mantovani, M. S., Bellini, M. F., Angeli, J. P. F., Oliveira, R. J., Silva, A. F., & Ribeiro, L. R. (2008). β-Glucans in promoting health: Prevention against mutation and cancer. *Mutation Research – Reviews in Mutation Research, 658*(3), 154–161. https://doi.org/10.1016/j.mrrev.2007.07.002

Mathews, R., Shete, V., & Chu, Y. F. (2021). The effect of cereal B-glucan on body weight and adiposity: A review of efficacy and mechanism of action. *Critical Reviews in Food Science and Nutrition, 0*(0), 1–13. https://doi.org/10.1080/10408398.2021.1994523

Mio, K., Yamanaka, C., Matsuoka, T., Kobayashi, T., & Aoe, S. (2020). Effects of β-glucan rich barley flour on glucose and lipid metabolism in the ileum, liver, and adipose tissues of high-fat diet induced-obesity model male mice analyzed by DNA microarray. *Nutrients, 12*(11), 1–18. https://doi.org/10.3390/nu12113546

Mirończuk-Chodakowska, I., Kujawowicz, K., & Witkowska, A. M. (2021). Beta-glucans from fungi: Biological and health-promoting potential in the covid-19 pandemic era. *Nutrients, 13*(11). https://doi.org/10.3390/nu13113960

Mishra, N. (2020). Cereal β glucan as a functional ingredient. *Innovations in Food Technology*, 109–122. https://doi.org/10.1007/978-981-15-6121-4_8

Miyamoto, J., Watanabe, K., Taira, S., Kasubuchi, M., Li, X., Irie, J., Itoh, H., & Kimura, I. (2018). Barley β-glucan improves metabolic condition via short-chain fatty acids produced by gut microbial fermentation in high fat diet fed mice. *PLoS One, 13*(4), 1–13. https://doi.org/10.1371/journal.pone.0196579

Mo, X., Sun, Y., Liang, X., Li, L., Hu, S., Xu, Z., Liu, S., Zhang, Y., Li, X., & Liu, L. (2022). Insoluble yeast β-glucan attenuates high-fat diet-induced obesity by regulating gut microbiota and its metabolites. *Carbohydrate Polymers, 281*(December 2021). https://doi.org/10.1016/j.carbpol.2021.119046

Mohiuddin, S. F. (2022). Barley beta-glucan mini-review: Benefits for post prandial glucose, lipid metabolism, and immune modulation. *SSRN Electronic Journal*. https://doi.org/10.2139/ssrn.4001720

Motta, F., Gershwin, M. E., & Selmi, C. (2021). Mushrooms and immunity. *Journal of Autoimmunity, 117*(November 2020), 102576. https://doi.org/10.1016/j.jaut.2020.102576

Murphy, E. J., Masterson, C., Rezoagli, E., O'Toole, D., Major, I., Stack, G. D., Lynch, M., Laffey, J. G., & Rowan, N. J. (2020a). β-glucan extracts from the same edible shiitake mushroom

Lentinus edodes produce differential in-vitro immunomodulatory and pulmonary cytoprotective effects—Implications for coronavirus disease (COVID-19) immunotherapies. *Science of the Total Environment, 732*, 139330. https://doi.org/10.1016/j.scitotenv.2020.139330

Murphy, E. J., Rezoagli, E., Major, I., Rowan, N. J., & Laffey, J. G. (2020b). B-glucan metabolic and immunomodulatory properties and potential for clinical application. *Journal of Fungi, 6*(4), 1–33. https://doi.org/10.3390/jof6040356

Najm, W. I. (2012). An overview on nutraceuticals and herbal supplements for diabetes and metabolic syndrome. *Nutritional and Therapeutic Interventions for Diabetes and Metabolic Syndrome*, 355–365. https://doi.org/10.1016/B978-0-12-385083-6.00028-0

Nehmi-Filho, V., Santamarina, A. B., de Freitas, J. A., Trarbach, E. B., de Oliveira, D. R., Palace-Berl, F., de Souza, E., de Miranda, D. A., Escamilla-Garcia, A., Otoch, J. P., & Pessoa, A. F. M. (2023). Novel nutraceutical supplements with yeast β-glucan, prebiotics, minerals, and Silybum marianum (silymarin) ameliorate obesity-related metabolic and clinical parameters: A double-blind randomized trial. *Frontiers in Endocrinology, 13*(January), 1–11. https://doi.org/10.3389/fendo.2022.1089938

Nieman, D. C., Henson, D. A., McMahon, M., Wrieden, J. L., Davis, J. M., Murphy, E. A., Gross, S. J., Mcanulty, L. S., & Dumke, C. L. (2008). β-Glucan, immune function, and upper respiratory tract infections in athletes. *Medicine and Science in Sports and Exercise, 40*(8), 1463–1471. https://doi.org/10.1249/MSS.0b013e31817057c2

Noronha, J. C., Zurbau, A., & Wolever, T. M. S. (2022). The importance of molecular weight in determining the minimum dose of oat β-glucan required to reduce the glycaemic response in healthy subjects without diabetes: A systematic review and meta-regression analysis. *European Journal of Clinical Nutrition, December 2021*. https://doi.org/10.1038/s41430-022-01176-5

Novak, M., & Vetvicka, V. (2008). β-glucans, history, and the present: Immunomodulatory aspects and mechanisms of action. *Journal of Immunotoxicology, 5*(1), 47–57. https://doi.org/10.1080/15476910802019045

Othman, R. A., Moghadasian, M. H., & Jones, P. J. H. (2011). Cholesterol-lowering effects of oat β-glucan. *Nutrition Reviews, 69*(6), 299–309. https://doi.org/10.1111/j.1753-4887.2011.00401.x

Pan, W., Jiang, P., Zhao, J., Shi, H., Zhang, P., Yang, X., Biazik, J., Hu, M., Hua, H., Ge, X., Huang, X. F., & Yu, Y. (2021). β-Glucan from Lentinula edodes prevents cognitive impairments in high-fat diet-induced obese mice: involvement of colon-brain axis. *Journal of Translational Medicine, 19*(1), 1–17. https://doi.org/10.1186/s12967-021-02724-6

Pelizon, A. C., Kaneno, R., Soares, A. M. V. C., Meira, D. A., & Sartori, A. (2005). Immunomodulatory activities associated with β-glucan derived from Saccharomyces cerevisiae. *Physiological Research, 54*(5), 557–564. https://doi.org/10.33549/physiolres.930686

Petravić-tominac, V., & Panjkota-krbavčić, I. (2010). Biological effects of yeast β-glucans. *Agriculturae conspectus scientificus, 75*(4), 149–158.

Priyanka, R., Sumana, C., & Prerona, S. (2020). Immunomodulatory Activity of Natural Polysaccharides in Combating Covid -19, Cancer, Inflammatory Disorders: A Review. *International Journal of Pharma and Bio Sciences, 10*(5), 191–206. https://doi.org/10.22376/ijpbs/lpr.2020.10.5.p191-206

Queenan, K. M., Stewart, M. L., Smith, K. N., Thomas, W., Fulcher, R. G., & Slavin, J. L. (2007). Concentrated oat β-glucan, a fermentable fiber, lowers serum cholesterol in hypercholesterolemic adults in a randomized controlled trial. *Nutrition Journal, 6*, 1–8. https://doi.org/10.1186/1475-2891-6-6

Rahmani, J., Miri, A., Černevičiūtė, R., Thompson, J., de Souza, N. N., Sultana, R., Kord Varkaneh, H., Mousavi, S. M., & Hekmatdoost, A. (2019). Effects of cereal beta-glucan consumption on body weight, body mass index, waist circumference and total energy intake: A meta-analysis of randomized controlled trials. *Complementary Therapies in Medicine, 43*, 131–139. https://doi.org/10.1016/j.ctim.2019.01.018

Redmond, M. (2005). *m a g a z i n e The global publication of the international federation of societies of cosmetic chemists anti-wrinkle therapy : significant new findings in the non-invasive cosmetic treatment of skin wrinkles with. March 2018*. https://doi.org/10.1111/j.1463-1318.2005.00268

Reyna-Villasmil, N., Bermúdez-Pirela, V., Mengual-Moreno, E., Arias, N., Cano-Ponce, C., Leal-Gonzalez, E., Souki, A., Inglett, G. E., Israili, Z. H., Hernández-Hernández, R., Valasco, M., & Arraiz, N. (2007). Oat-derived β-glucan significantly improves HDLC and diminishes LDLC and non-HDL cholesterol in overweight individuals with mild hypercholesterolemia. *American Journal of Therapeutics, 14*(2), 203–212. https://doi.org/10.1097/01.pap.0000249917.96509.e7

Rodrigues, M. V., Zanuzzo, F. S., Koch, J. F. A., de Oliveira, C. A. F., Sima, P., & Vetvicka, V. (2020). Development of Fish Immunity and the Role of β-Glucan in Immune Responses. *Molecules, 25*(22), 1–33. https://doi.org/10.3390/MOLECULES25225378

Savelkoul, H. F. J., Chanput, W., & Wichers, H. J. (2013). Immunomodulatory effects of mushroom β-glucans. In *Diet, Immunity and Inflammation*. Woodhead Publishing Limited. https://doi.org/10.1533/9780857095749.3.416

Schlörmann, W., Atanasov, J., Lorkowski, S., Dawczynski, C., & Glei, M. (2020). Study on chemopreventive effects of raw and roasted β-glucan-rich waxy winter barley using an: In vitro human colon digestion model. *Food and Function, 11*(3), 2626–2638. https://doi.org/10.1039/c9fo03009c

Sengül, M., & Ufuk, S. (2022). Therapeutic and functional properties of beta-glucan and its effects on health. *Eurasian Journal of Food Science and Technology, 6*(1), 29–41.

Shehzad, A., Rabail, R., Munir, S., Jan, H., Fernández-Lázaro, D., & Aadil, R. M. (2023). Impact of oats on appetite hormones and body weight management: A review. *Current Nutrition Reports, 12*(1), 66–82. https://doi.org/10.1007/s13668-023-00454-3

Shen, R. L., Dang, X. Y., Dong, J. L., & Hu, X. Z. (2012). Effects of oat β-glucan and barley β-glucan on fecal characteristics, intestinal microflora, and intestinal bacterial metabolites in rats. *Journal of Agricultural and Food Chemistry, 60*(45), 11301–11308. https://doi.org/10.1021/jf302824h

Shimizu, K. (2014, June). *Efficacy of oral administered superfine dispersed lentinan for advanced pancreatic cancer*.

Shoukat, M., & Sorrentino, A. (2021). Cereal β-glucan: a promising prebiotic polysaccharide and its impact on the gut health. *International Journal of Food Science and Technology, 56*(5), 2088–2097. https://doi.org/10.1111/ijfs.14971

Sima, P., Vannucci, L., & Vetvicka, V. (2018). β-glucans and cholesterol (Review). *International Journal of Molecular Medicine, 41*(4), 1799–1808. https://doi.org/10.3892/ijmm.2018.3411

Singh, R. P., & Bhardwaj, A. (2023). B-Glucans: A potential source for maintaining gut microbiota and the immune system. *Frontiers in Nutrition, 10*(May), 1–22. https://doi.org/10.3389/fnut.2023.1143682

Sivieri, K., de Oliveira, S. M., Marquez, A. d. S., Pérez-Jiménez, J., & Diniz, S. N. (2022). Insights on β-glucan as a prebiotic coadjuvant in the treatment of diabetes mellitus: A review. *Food Hydrocolloids for Health, 2*(January). https://doi.org/10.1016/j.fhfh.2022.100056

Smith, K. N., Queenan, K. M., Fulcher, R. G., Slavin, J. L., & Thomas, W. (2008). Physiological effects of concentrated barley β-Glucan in mildly hypercholesterolemic adults. *Journal of the American College of Nutrition, 27*(3), 434–440. https://doi.org/10.1080/07315724.2008.10719722

Suchecka, D., Gromadzka-Ostrowska, J., Zyła, E., Harasym, J. P., & Oczkowski, M. (2017). Selected physiological activities and health promoting properties of cereal beta-glucans. A review. *Journal of Animal and Feed Sciences, 26*(3), 183–191. https://doi.org/10.22358/jafs/70066/2017

Tappy, L., Gügolz, E., & Würsch, P. (1996). Effects of breakfast cereals containing various amounts of β-glucan fibers on plasma glucose and insulin responses in NIDDM subjects. *Diabetes Care, 19*(8), 831–834. https://doi.org/10.2337/diacare.19.8.831

Therdtatha, P., Shinoda, A., & Nakayama, J. (2022). Crisis of the Asian gut: Associations among diet, microbiota, and metabolic diseases. *Bioscience of Microbiota, Food and Health, 41*(3), 83–93. https://doi.org/10.12938/bmfh.2021-085

Thomas, S., Rezoagli, E., Abidin, I. Z., Major, I., Murray, P., & Murphy, E. J. (2022). β-Glucans from Yeast—Immunomodulators from novel waste resources. *Applied Sciences (Switzerland), 12*(10). https://doi.org/10.3390/app12105208

Thompson, I. J., Oyston, P. C. F., & Williamson, D. E. (2010). Potential of the β-glucans to enhance innate resistance to biological agents. *Expert Review of Anti-Infective Therapy, 8*(3), 339–352. https://doi.org/10.1586/eri.10.10

Titus Sombuor. (2022). Anti-proliferative effects of lentinan, a βeta-glucan from shiitake mushroom-Titus. 2005–2003 ,*8.5.2017* ,*הארץ*. www.aging-us.com

Tiwari, U., & Cummins, E. (2009). Factors influencing β-glucan levels and molecular weight in cereal-based products. *Cereal Chemistry, 86*(3), 290–301. https://doi.org/10.1094/CCHEM-86-3-0290

Van Steenwijk, H. P., & Bast, A. (2021). Immunomodulating effects of fungal beta-glucans. *Nutrients, 13*, 1–20.

Velikonja, A., Lipoglavšek, L., Zorec, M., Orel, R., & Avguštin, G. (2019). Alterations in gut microbiota composition and metabolic parameters after dietary intervention with barley beta glucans in patients with high risk for metabolic syndrome development. *Anaerobe, 55*, 67–77. https://doi.org/10.1016/j.anaerobe.2018.11.002

Vetvicka, V. (2011). Glucan-immunostimulant, adjuvant, potential drug. *World Journal of Clinical Oncology, 2*(2), 115. https://doi.org/10.5306/wjco.v2.i2.115

Vetvicka, V., Vannucci, L., Sima, P., & Richter, J. (2019). Beta glucan: Supplement or drug? From laboratory to clinical trials. *Molecules, 24*(7). https://doi.org/10.3390/molecules24071251

Vetvicka, V., Vannucci, L., & Sima, P. (2020). B-Glucan as a new tool in vaccine development. *Scandinavian Journal of Immunology, 91*(2), 0–2. https://doi.org/10.1111/sji.12833

Volman, J. J., Ramakers, J. D., & Plat, J. (2008). Dietary modulation of immune function by β-glucans. *Physiology and Behavior, 94*(2), 276–284. https://doi.org/10.1016/j.physbeh.2007.11.045

Wang, Y., Ames, N. P., Tun, H. M., Tosh, S. M., Jones, P. J., & Khafipour, E. (2016). High molecular weight barley β-glucan alters gut microbiota toward reduced cardiovascular disease risk. *Frontiers in Microbiology, 7*(FEB), 1–15. https://doi.org/10.3389/fmicb.2016.00129

Wang, H., Chen, G., Li, X., Zheng, F., & Zeng, X. (2020). Yeast β-glucan, a potential prebiotic, showed a similar probiotic activity to inulin. *Food and Function, 11*(12), 10386–10396. https://doi.org/10.1039/d0fo02224a

Wani, S. M., Gani, A., Mir, S. A., Masoodi, F. A., & Khanday, F. A. (2021). β-Glucan: A dual regulator of apoptosis and cell proliferation. *International Journal of Biological Macromolecules, 182*, 1229–1237. https://doi.org/10.1016/j.ijbiomac.2021.05.065

Wolever, T., Rahn, M., Dioum, E., Spruill, S. E., Ezatagha, A., Campbell, J. E., Jenkins, A. L., & Chu, Y. (2021). An Oat β-Glucan beverage reduces LDL cholesterol and cardiovascular disease risk in men and women with Borderline high cholesterol: A double-blind, randomized, controlled clinical trial. *Journal of Nutrition, 151*(9), 2655–2666. https://doi.org/10.1093/jn/nxab154

Wood, P. J. (2010). Oat and rye β-glucan: Properties and function. *Cereal Chemistry, 87*(4), 315–330. https://doi.org/10.1094/CCHEM-87-4-0315

Wouk, J., Dekker, R. F. H., Queiroz, E. A. I. F., & Barbosa-Dekker, A. M. (2021). β-Glucans as a panacea for a healthy heart? Their roles in preventing and treating cardiovascular diseases. *International Journal of Biological Macromolecules, 177*, 176–203. https://doi.org/10.1016/j.ijbiomac.2021.02.087

Wu, Z., Ming, J., Gao, R., Wang, Y., Liang, Q., Yu, H., & Zhao, G. (2011). Characterization and antioxidant activity of the complex of tea polyphenols and oat β-glucan. *Journal of Agricultural and Food Chemistry, 59*(19), 10737–10746. https://doi.org/10.1021/jf202722w

Würsch, P., & Pi-Sunyer, F. X. (1997). The role of viscous soluble fiber in the metabolic control of diabetes: A review with special emphasis on cereals rich in β-glucan. *Diabetes Care, 20*(11), 1774–1780. https://doi.org/10.2337/diacare.20.11.1774

Xin-Zhong, H., Xia-Lu, S., Xiao-Ping, L., Liu, L., Jian-Mei, Z., & Xing-Yun, C. (2015). Effect of dietary oat β-glucan on high-fat diet induced obesity in HFA mice. *Bioactive Carbohydrates and Dietary Fibre, 5*(1), 79–85. https://doi.org/10.1016/j.bcdf.2014.12.006

Yoo, J. Y., & Kim, S. S. (2016). Probiotics and prebiotics: Present status and future perspectives on metabolic disorders. *Nutrients, 8*(3), 1–20. https://doi.org/10.3390/nu8030173

Yu, C., Chen, H., Du, D., Lv, W., Li, S., Li, D., Xu, Z., Gao, M., Hu, H., & Liu, D. (2021). β-Glucan from Saccharomyces cerevisiae alleviates oxidative stress in LPS-stimulated RAW264.7 cells via Dectin-1/Nrf2/HO-1 signaling pathway. *Cell Stress and Chaperones, 26*(4), 629–637. https://doi.org/10.1007/s12192-021-01205-5

Yu, L., Gao, Y., Ye, Z., Duan, H., Zhao, J., Zhang, H., Narbad, A., Tian, F., Zhai, Q., & Chen, W. (2023). Interaction of beta-glucans with gut microbiota: Dietary origins, structures, degradation, metabolism, and beneficial function. *Critical Reviews in Food Science and Nutrition.* https://doi.org/10.1080/10408398.2023.2217727

Yun, C. H., Estrada, A., Van Kessel, A., Park, B. C., & Laarveld, B. (2003). β-Glucan, extracted from oat, enhances disease resistance against bacterial and parasitic infections. *FEMS Immunology and Medical Microbiology, 35*(1), 67–75. https://doi.org/10.1016/S0928-8244(02)00460-1

Zaremba, S. M. M., Gow, I. F., Drummond, S., McCluskey, J. T., & Steinert, R. E. (2018). Effects of oat β-glucan consumption at breakfast on ad libitum eating, appetite, glycemia, insulinemia and GLP-1 concentrations in healthy subjects. *Appetite, 128*(March), 197–204. https://doi.org/10.1016/j.appet.2018.06.019

Zeković, D. B., Kwiatkowski, S., Vrvić, M. M., Jakovljević, D., & Moran, C. A. (2005). Natural and modified (1→3)-β-D-glucans in health promotion and disease alleviation. *Critical Reviews in Biotechnology, 25*(4), 205–230. https://doi.org/10.1080/07388550500376166

Zhang, M., Cheung, P. C. K., Zhang, L., Chiu, C. M., & Ooi, V. E. C. (2004). Carboxymethylated β-glucans from mushroom sclerotium of Pleurotus tuber-regium as novel water-soluble anti-tumor agent. *Carbohydrate Polymers, 57*(3), 319–325. https://doi.org/10.1016/j.carbpol.2004.05.008

Zhao, L., Dong, Y., Chen, G., & Hu, Q. (2010). Extraction, purification, characterization and antitumor activity of polysaccharides from Ganoderma lucidum. *Carbohydrate Polymers, 80*(3), 783–789. https://doi.org/10.1016/j.carbpol.2009.12.029

Zhonghui, Z., Xiaowei, Z., & Fang, F. (2014). Ganoderma lucidum polysaccharides supplementation attenuates exercise-induced oxidative stress in skeletal muscle of mice. *Saudi Journal of Biological Sciences, 21*(2), 119–123. https://doi.org/10.1016/j.sjbs.2013.04.004

Zhou, X., Luo, L., Dressel, W., Shadier, G., Krumbiegel, D., Schmidtke, P., Zepp, F., & Meyer, C. U. (2008). Cordycepin is an immunoregulatory active ingredient of cordyceps sinensis. *American Journal of Chinese Medicine, 36*(5), 967–980. https://doi.org/10.1142/S0192415X08006387

# Chapter 10
# Stability of Beta-Glucans During Processing

**Mudasir Bashir Mir and Shabir Ahmad Mir**

## 10.1 Introduction

Beta glucan (β-glucan) is a soluble dietary fiber obtained from oats, barley and other sources. β-glucans are a heterogeneous group of non-starch polysaccharides, consisting of D-glucose monomers linked by β-glycosidic bonds (Zeković et al., 2005). The macromolecular structure of β-glucan depends on both the source and method of isolation. The simplest glucan is the linear and unbranched β-(1,3)-D-glucan (McIntosh et al., 2005). Glucans from barley, oats, or wheat are found in cell walls of the endosperm, while being concentrated in the aleurone layer of barley, oats, wheat, sorghum, and other cereals. Branched structures of β-glucans consist of β-(1,3)- or β-(1,4)-glucan backbone with either (1,2)- or (1,6)-linked β-glucopyranosyl side branches (Barsanti et al., 2011). They are major structural components of the cell walls of yeast, fungi, and some bacteria (Volman et al., 2008). Branched β-(1,4;1,6)-D-glucan and β-(1,3;1,6)-D-glucan are found in different groups of yeast, fungi, and algae (Barsanti et al., 2011). Besides differences in type of linkage and branching, β-glucans can vary in terms of frequency and length of branching, degree of branching, molecular weight (from $10^2$ to $10^6$ daltons), polymer charge, and/or solution conformation (random coil or triple or single helix) as well as solubility (Soltanian et al., 2009). All these factors play a role in shaping β-glucan-associated biological activities. The molecular complexity of β-glucan, characterized by unique β-glycosidic linkages, sets it apart from other polysaccharides and underlies its multifaceted properties (Jayachandran et al., 2018).

---

M. Bashir Mir
Department of Food Science & Technology, Govt. Degree College,
Baramulla, Jammu & Kashmir, India

S. A. Mir (✉)
Department of Food Science & Technology, Govt. College for Women,
M. A. Road, Srinagar, Jammu & Kashmir, India

 
M. Bashir Mir, S. A. Mir (eds.), *Beta-Glucan: Sources, Properties and Applications*,
https://doi.org/10.1007/978-3-031-95788-8_10

It is used as a functional ingredient and a means to improve dietary fiber content in foods. It is gaining more importance, and finds spectrum of applications in food and allied industries owing to its numerous health benefits. Many of the health benefits, including cholesterol lowering ability, hypoglycemic, and laxative effects of beta-glucan are attributed to its high viscosity, molecular weight and fermentability (Brummer et al., 2012; Regand et al., 2009; Tosh et al., 2010; Wolever et al., 2010; El Khoury et al., 2012). The physicochemical properties such as molecular weight, structural confirmations etc. are the major determinants of their changes in viscosity (Henrion et al., 2019). Therefore, it becomes imperative to study these properties in detail while quantifying their impact on physiological responses. β-glucan rich cereal grains such as oats and barley are rarely consumed raw. They need to be cooked before being eaten safely. Cooking and other processing conditions will generally modify physicochemical characteristics of β-glucans, such as molecular weight, extractability and viscosity. The present chapter attempts to study the impact of cooking and other processing treatments on various properties.

## 10.2 Effect of Drying

The effect of drying was studied on different properties of oat β-glucan extracts such as gelation, water, retention capacity displacement etc. Drying of oat β-glucan extracts were carried at the temperatures of 25 °C, 50 °C, 75 °C and 100 °C in order to decrease the moisture content to desired levels and it was observed that no significant impact was observed in water retention capacity of β-glucan at temperatures of 25 °C & 50 °C (Oliveira et al., 2012). However, extract samples dried at 75 °C and 100 °C showed significant reduction in water retention capacity. β-glucan content was found to be reduced in oat grains at higher temperatures, the lowest value being at the temperature of 100 °C (Kaur et al., 2019). The high temperatures used in drying may have caused the degradation β-glucan into fragments or it may even have depolymerized the linear structure (Butt et al., 2008). Gelation capacity and displacement of β-glucan extract were found to increase with increase in air drying temperature, the highest being reported at a temperature of 100 °C (Oliveira et al., 2012).

## 10.3 Effect of Milling

The process of milling has been found to make significant impact on the yield, purity and properties like solubility, viscosity etc. of extracted β-glucan. The novel method of freeze-milling improved the yield in addition to the enhancement of purity of β-glucan obtained. After the extraction from milled particles, purity of the achieved beta-glucan preparation was 84.4% when compared to the control where purity was 74.1% (Jurkaninová et al., 2024). Harasym et al. (2015) showed that freeze-milling process reduced the particle size of the β-glucan significantly. Increased solubility with a significant decline in viscosity of β-glucan was attributed to the low molecular weight of the extracted β-glucan. Izydorczyk and Dexter (2008) reported that milling process can also be influenced by different milling equipment. Intensive mechanical processing, as identified by Wood et al. (1989), may result in shearing damage to β-D-glucans, causing changes in their physical properties such as solubility and viscosity. No significant effect of milling and pearling was observed on the β-D-glucan content of barley (KeShun et al., 2009; Sullivan et al., 2010).

## 10.4 Effect of Germination

The Sprouting of barley is done with a purpose to improve its functionality and flavor development in brewing / distilling industries. β-D-glucan content of barley during germination was significantly reduced due to the action of an endogenous enzyme, β-glucanase. Nielsen and Munck (2003) and Wang and Zhao (2007) studied the various properties of sprouted oats and concluded that β-D-glucan content declined due to the activation of β-D-glucan degrading enzyme but the decrease in β-D-glucans is initially very slow (Wilhelmson et al., 2001). Barley varieties that contain high amount of beta-D-glucan content showed gradual decrease in the content after 24 h germination period. However, the decrease in beta-D-glucan content became very significant after 96 h of germination period (Islam et al., 2021). It can therefore be concluded that short sprouting times are recommended in order to retain the soluble fiber and beta-D-glucan content of oats and barley (Hübner et al., 2010; Jurkaninová et al., 2024). Higher content of β-D-glucans in malt is undesirable in brewing which is attributed to the insufficient degradation of cell wall and hindering the diffusion germination enzymes (Etokakpan, 1993; Wang et al., 2004). Better malting performance is therefore positively correlated to lower levels β-D-glucans and higher levels of β-glucanases, and to achieve the desired β-D-glucans content in barley malt (100–300 mg.$L^{-1}$), have been studied in detail in the brewing industry (Wang et al., 2004; Edison et al., 2022).

## 10.5 Effect of Cooking

Different studies have been conducted to see the effect of cooking on β-D-glucans from different sources. Morales et al. (2024) studied the impact of various household cooking treatments on β-glucan content of White-button (*Agaricus bisporus*) and shiitake (*Lentinula edodes*) mushrooms. A lot of discrepancy was observed in the amount of β-glucan depending upon the cooking treatment. The study demonstrated that frying treatment (120–180 °C) to both mushrooms was detrimental to their stability and caused drastic reduction in the concentration of β-glucan content in comparison to their raw counterparts. However, boiling treatment led to enhancement in the levels of β-glucan content unlike other treatments. Yiu et al. (1987) studied the influence of cooking on β-glucan of rolled oats and observed that gradually cooked oatmeal had more solubilized (1,3)/(1,4)- β -D-glucan and starch and had a more viscous supernatant. Barley varieties subjected to autoclaving coupled with the use of different chemicals such as calcium chloride, ethanol, hydrochloric acid, and sodium hydroxide was found to preserve β-glucan content by hindering the activity of β-glucanase (Knuckles & Chiu, 1999). Zhang (2010) while working on β-D-glucans from edible fungus (*Pleurotus geestanus*) reported two transition breakdown of purified β-D-glucans following heating in water. The treatment apparently changed the conformation and viscosity of β-D-glucans.

## 10.6 Effect of Baking

Several studies described the impact of baking on β-glucan content of bread. The effect was evaluated in terms of its degradation, molecular weight changes, and extractability. β- glucan is incorporated in breads as a functional ingredient and a means to improve the dietary fiber content. However, the fate of β-glucans depends on the type of food processing method employed. Lu et al. (2019) studied the effect of controlled dough fermentation on the content, molecular weight (MW), and viscosity of β-D-glucans. The total β-glucan content decreased after 12 h fermentation. The activities of lactic acid bacteria (*Lactobacillus plantarum*) and β-glucanase are responsible for such a decrease in β-glucan content. During bread making, the endogenous β-glucanase present in cereal flour degraded β-glucans but remain unaffected by the presence of yeast (Andersson et al., 2004). A decrease was also observed in the MW of β-D-glucans and their content in barley and rye dough with a fermentation time of 0–60 min, and in a sourdough from barley whole meal flour and oat bran leaven under fermentation conditions of 30 °C for 20 h (Andersson et al., 2009; Andersson & Åman, 2000; Rieder et al., 2012). Some studies highlighted no significant changes in β-glucans in yeast-raised rye bread but a slight increase in solubility was noted (Comino et al., 2016). Similarly, Maina et al. (2021) didn't find any discernible effect of baking on the molecular structure and total content of β-glucans. Short fermentation times favor increase in β-glucans due to improved extractability of β-glucan polymers through degradation of other cell wall polysaccharides that are naturally entangled with β-glucans in the grain (Andersson et al., 2008; Henrion et al., 2019). Further, the scope of research, with regard to the effect of baking on β-glucans, need to be widened to fully understand their fate.

## 10.7 Effect of Extrusion

Extrusion process is one of the important unit operation used for the preparation of extrudates. Extrusion process involved many unit operations including, mixing, shearing, heating, kneading, shaping and forming. Extrusion process improves the technological properties of materials from different industries, useful to develop attractive ingredients with enhanced nutritional and technological characteristics. Beta-glucan content was observed highest in cereal flours as compared to their extrudates. The reduction in beta-glucan content is due to the high shear extrusion conditions during the process. However, some studies also reported that beta-glucan in extruded cereals is not affected by extrusion conditions such as change in extrusion temperature, and moisture content (Yao et al., 2006). The soluble β-D glucans increase in barley after extrusion cooking with the highest observed in extrudates prepared at high temperature and low moisture conditions (Sharma & Gujral, 2013). In control barley, the ratio of soluble to insoluble β-D glucans ranged from 0.7 to 1.5, however, after extrusion process, the ratio changed from 1.2 to 3.1. The increase in level is due to the extrusion cooking which leads to the changes in the molecular mass and solubility of β-D glucans. Furthermore, the extrusion conditions lead to depolymerizaiton, disintegration of cell walls which affect the distribution of β-D glucans in the cereals (Tosh et al., 2010).

## 10.8 Effect of Roasting

Roasting is one of the simple unit operations that used dry heat for improving the properties of food. Roasting leads to the gelatinization of starch and denaturation of proteins thus improves the digestibility of grains and also increased the resistant starch content. Roasting process significantly increased the insoluble beta-glucan with a concomitant decrease in the soluble beta-glucan in barley. Heat treatment decreased the beta-glucan content of oat flour after roasting process. The high temperature during roasting may lead to the depolymerisation and degradation of beta-glucan, resulting in the destruction of beta-glucan by unwinding the molecular chain (Schlörmann et al., 2020). The effect of thermal processing on β-glucan varied with time-temperature combination and thermal properties. Furthermore, the decrease of beta-glucan solubility during heating process also leads to the reduction of beta-glucan content (Sharma et al., 2011).

## 10.9 Effect of Low Temperature

Low temperature process reduces the molecular weight and solubility of beta-glucan. Lan-Pidhainy et al. (2007) observed that freezing process decreased the molecular weight and solubility of beta-glucan with increase of freeze-thaw cycles.

Beer et al. (1997) observed that under frozen storage at −20 °C for 5 months, extractable β-glucan decreased by more than 50% in oat bran muffins. Beta-glucan properties including solubility, viscosity and molecular weight in fresh oat bran bread remain unchanged for first 3 days of storage period at room temperature, and then reduction in the values of these properties was observed. The freeze drying of oat bran bread affects the beta-glucan solubility and viscosity by removing the water from beta-glucan, protein, and starch (Gamel et al., 2013). The storage conditions, including freezing at −20 °C, affected the solubility and viscosity of β-glucan extracted from bread fortified with barley β-glucan (Moriartey et al., 2011).

## 10.10 Mechanical Processing

Many mechanical processes are used to reduce the molecular weight of beta-glucan and ultimately affect the functionality of beta-glucan. The mechanical process applied to beta-glucan food leads to the breakdown of glycosidic linkages in the beta-glucan structure which results in the reduction of molecular weight, which further leads to the water solubility and other physiological functions.

The effect of homogenization on the properties of beta-glucan was investigated by Kivelä et al. (2010). Homegenization at high pressure modified the beta-glucan during storage. The results showed the decline in molecular weight, viscosity and shear thinning behaviour of beta-glucan based samples due to high pressure homogenization. The decrease of molecular weight might be due to the chain length breakdown of beta-glucan linkages. Homogenization has also shown the effect on properties of beta-glucan from yeast, *Saccharomyces cerevisiae*. The results showed the higher beta-glucan yield after homogenization with higher apparent viscosity as compared to untreated yeast (Thammakiti et al., 2004). The homogenization process leads to the fragmentation of cell walls and hence improved the release of beta-glucan from yeast cells.

## 10.11 Chemical Methods

Chemical treatments are also used to modify the beta-glucans. Chemical agents including acid, alkali, and salth has the ability to break the glycosidic bonds in beta-glucan leading to change in viscosity and molecular weight. The decreased in viscosity and molecular weight of cereal beta-glucan were observed by Kivelä et al. (2009) after treating with ascorbic acid. The chemical modification is primarily controlled the presence of beta-glucan that acts as a point for the attachment of other functional groups. The attachment of functional groups to the beta-glucans provides providing modification forms to the beta-glucans. Different studies have reported the modification of beta-glucans due to the utilization of different chemical agents in beta-glucan containing foods (Wang et al., 2005).

## 10.12 Conclusion

β-D-glucan content of barley reduced significantly during germination. Prolonged fermentation decreased the total β-glucan content. Boiling treatment led to enhancement in the levels of β-glucan content unlike other treatments. The soluble β-D glucans increase in barley after extrusion cooking at high temperature and low moisture conditions. Roasting significantly increased the insoluble beta-glucan in barley. Low temperature processing reduces the molecular weight and solubility of beta-glucan.

## References


Andersson, R., & Åman, P. (2000). Cereal arabinoxylan: Occurrence, structure and properties. In *Advanced dietary fibre technology* (pp. 299–314). Blackwell Science.

Andersson, A. A., Armö, E., Grangeon, E., Fredriksson, H., Andersson, R., & Åman, P. (2004). Molecular weight and structure units of (1→ 3, 1→ 4)-β-glucans in dough and bread made from hull-less barley milling fractions. *Journal of Cereal Science, 40*(3), 195–204.

Andersson, A. A. M., Rüegg, N., & Åman, P. (2008). Molecular weight distribution and content of water-extractable β-glucan in rye crisp bread. *Journal of Cereal Science, 47*(3), 399–406.

Andersson, R., Fransson, G., Tietjen, M., & Åman, P. (2009). Content and molecular-weight distribution of dietary fiber components in whole-grain rye flour and bread. *Journal of Agricultural and Food Chemistry, 57*(5), 2004–2008.

Barsanti, L., Passarelli, V., Evangelista, V., Frassanito, A. M., & Gualtieri, P. (2011). Chemistry, physico-chemistry and applications linked to biological activities of β-glucans. *Natural Product Reports, 28*(3), 457–466.

Beer, M. U., Wood, P. J., Weisz, J., & Fillion, N. (1997). Effect of cooking and storage on the amount and molecular weight of (1→ 3)(1→ 4)-β-D-glucan extracted from oat products by an in vitro digestion system. *Cereal Chemistry, 74*(6), 705–709.

Brummer, Y., Duss, R., Wolever, T. M. S., & Tosh, S. M. (2012). Glycemic response to extruded oat bran cereals processed to vary in molecular weight. *Cereal Chemistry, 89*(5), 255–261.

Butt, M. S., Tahir-Nadeem, M., Khan, M. K. I., Shabir, R., & Butt, M. S. (2008). Oat: Unique among the cereals. *European Journal of Nutrition, 47*, 68–79.

Comino, P., Collins, H., Lahnstein, J., & Gidley, M. J. (2016). Effects of diverse food processing conditions on the structure and solubility of wheat, barley and rye endosperm dietary fibre. *Journal of Food Engineering, 169*, 228–237.

Edison, L. K., Reji, S. R., & Pradeep, N. S. (2022). Beta-glucanase in breweries. In *Microbial beta glucanases: Molecular structure, functions and applications* (pp. 85–98). Singapore.

El Khoury, D., Cuda, C., Luhovyy, B. L., & Anderson, G. H. (2012). Beta glucan: Health benefits in obesity and metabolic syndrome. *Journal of Nutrition and Metabolism, 2012*(1), 851362.

Etokakpan, O. U. (1993). Enzymatic degradation and nature of the endosperm cell-walls of germinating sorghums and barley. *Journal of Science, Food and Agriculture, 61*, 389–393.

Gamel, T. H., Badali, K., & Tosh, S. M. (2013). Changes of β-glucan physicochemical characteristics in frozen and freeze dried oat bran bread and porridge. *Journal of Cereal Science, 58*(1), 104–109.

Harasym, J., Suchecka, D., & Gromadzka-Ostrowska, J. (2015). Effect of size reduction by freeze-milling on processing properties of beta-glucan oat bran. *Journal of Cereal Science, 61*, 119–125.

Henrion, M., Francey, C., Lê, K. A., & Lamothe, L. (2019). Cereal B-glucans: The impact of processing and how it affects physiological responses. *Nutrients, 11*(8), 1729.

Hübner, F., O'Neil, T., Cashman, K. D., & Arendt, E. K. (2010). The influence of germination conditions on beta-glucan, dietary fibre and phytate during the germination of oats and barley. *European Food Research and Technology, 231*, 27–35.

Islam, M. Z., An, H. G., Kang, S. J., & Lee, Y. T. (2021). Physicochemical and bioactive properties of a high β-glucan barley variety 'Betaone'affected by germination processing. *International Journal of Biological Macromolecules, 177*, 129–134.

Izydorczyk, M. S., & Dexter, J. E. (2008). Barley β-glucans and arabinoxylans: Molecular structure, physicochemical properties, and uses in food products–a review. *Food Research International, 41*(9), 850–868.

Jayachandran, M., Chen, J., Chung, S. S. M., & Xu, B. (2018). A critical review on the impacts of β-glucans on gut microbiota and human health. *The Journal of Nutritional Biochemistry, 61*, 101–110.

Jurkaninová, L., Dvořáček, V., Gregusová, V., & Havrlentová, M. (2024). Cereal β-d-Glucans in food processing applications and nanotechnology research. *Food, 13*(3), 500.

Kaur, R., Sharma, M., Ji, D., Xu, M., & Agyei, D. (2019). Structural features, modification, and functionalities of beta-glucan. *Fibers, 8*(1), 1.

KeShun, L., Barrows, F. T., & Don, O. (2009). Dry fractionation methods to produce barley meals varying in protein, beta-glucan, and starch contents. *Journal of Food Science, 74*, C487–C499.

Kivelä, R., Gates, F., & Sontag-Strohm, T. (2009). Degradation of cereal beta-glucan by ascorbic acid induced oxygen radicals. *Journal of Cereal Science, 49*(1), 1–3.

Kivelä, R., Pitkänen, L., Laine, P., Aseyev, V., & Sontag-Strohm, T. (2010). Influence of homogenisation on the solution properties of oat β-glucan. *Food Hydrocolloids, 24*(6–7), 611–618.

Knuckles, B. E., & Chiu, M. C. M. (1999). β-Glucanase activity and molecular weight of β-glucans in barley after various treatments. *Cereal Chemistry, 76*(1), 92–95.

Lan-Pidhainy, X., Brummer, Y., Tosh, S. M., Wolever, T. M., & Wood, P. J. (2007). Reducing beta-glucan solubility in oat bran muffins by freeze-thaw treatment attenuates its hypoglycemic effect. *Cereal Chemistry, 84*(5), 512–517.

Lu, J., Shan, L., Xie, Y., Min, F., Gao, J., Guo, L., et al. (2019). Effect of fermentation on content, molecule weight distribution and viscosity of β-glucans in oat sourdough. *International Journal of Food Science & Technology, 54*(1), 62–67.

Maina, N. H., Rieder, A., De Bondt, Y., Mäkelä-Salmi, N., Sahlstrøm, S., Mattila, O., et al. (2021). Process-induced changes in the quantity and characteristics of grain dietary fiber. *Food, 10*(11), 2566.

McIntosh, M., Stone, B. A., & Stanisich, V. A. (2005). Curdlan and other bacterial (1→ 3)-β-D-glucans. *Applied Microbiology and Biotechnology, 68*, 163–173.

Morales, D., Bal, M. A., Figueredo, S., Soler-Rivas, C., & Ruiz-Rodríguez, A. (2024). Effect of household cooking treatments on the stability of β-glucans, ergosterol, and phenolic compounds in white-button (Agaricus bisporus) and shiitake (Lentinula edodes) mushrooms. *Food and Bioprocess Technology, 17*(3), 791–798.

Moriartey, S., Temelli, F., & Vasanthan, T. (2011). Effect of storage conditions on the solubility and viscosity of β-glucan extracted from bread under in vitro conditions. *Journal of Food Science, 76*(1), C1–C7.

Nielsen, J. P., & Munck, L. (2003). Evaluation of malting barley quality using exploratory data analysis. I. Extraction of information from micro-malting data of spring and winter barley. *Journal of Cereal Science, 38*(2), 173–180.

Oliveira, L. D. C., Oliveira, M., Meneghetti, V. L., Mazzutti, S., Colla, L. M., Elias, M. C., & Gutkoski, L. C. (2012). Effect of drying temperature on quality of β-glucan in white oat grains. *Food Science and Technology, 32*, 775–783.

Regand, A., Tosh, S. M., Wolever, T. M. S., & Wood, P. J. (2009). Physicochemical properties of glucan in differently processed oat foods influence glycemic response. *Journal of Agricultural and Food Chemistry, 57*(19), 8831–8838.

Rieder, A., Holtekjølen, A. K., Sahlstrøm, S., & Moldestad, A. (2012). Effect of barley and oat flour types and sourdoughs on dough rheology and bread quality of composite wheat bread. *Journal of Cereal Science, 55*(1), 44–52.

Schlörmann, W., Zetzmann, S., Wiege, B., Haase, N. U., Greiling, A., Lorkowski, S., et al. (2020). Impact of different roasting conditions on sensory properties and health-related compounds of oat products. *Food Chemistry, 307*, 125548.

Sharma, P., & Gujral, H. S. (2013). Extrusion of hulled barley affecting β-glucan and properties of extrudates. *Food and Bioprocess Technology, 6*, 1374–1389.

Sharma, P., Gujral, H. S., & Rosell, C. M. (2011). Effects of roasting on barley β-glucan, thermal, textural and pasting properties. *Journal of Cereal Science, 53*(1), 25–30.

Soltanian, S., Stuyven, E., Cox, E., Sorgeloos, P., & Bossier, P. (2009). Beta-glucans as immunostimulant in vertebrates and invertebrates. *Critical Reviews in Microbiology, 35*(2), 109–138.

Sullivan, P., O'Flaherty, J., Brunton, N., Gee, V. L., Arendt, E., & Gallagher, E. (2010). Chemical composition and microstructure of milled barley fractions. *European Food Research and Technology, 230*(4), 579–595.

Thammakiti, S., Suphantharika, M., Phaesuwan, T., & Verduyn, C. (2004). Preparation of spent brewer's yeast β-glucans for potential applications in the food industry. *International Journal of Food Science & Technology, 39*(1), 21–29.

Tosh, S. M., Brummer, Y., Miller, S. S., Regand, A., Defelice, C., Duss, R., Wolever, T. M. S., & Wood, P. J. (2010). Processing affects the physicochemical properties of β-glucan in oat bran cereal. *Journal of Agricultural and Food Chemistry, 58*(13), 7723–7730.

Volman, J. J., Ramakers, J. D., & Plat, J. (2008). Dietary modulation of immune function by β-glucans. *Physiology & Behavior, 94*(2), 276–284.

Wang, N., & Zhao, M. (2007). Study progress on b-glucan in barley. *Science and Technology Qinghai Agriculture and Forestry, 3*, 34–36.

Wang, J., Zhang, G., Chen, J., & Wu, F. (2004). The changes of β-glucan content and β-glucanase activity in barley before and after malting and their relationships to malt qualities. *Food Chemistry, 86*(2), 223–228.

Wang, Y. J., Yao, S. J., Guan, Y. X., Wu, T. X., & Kennedy, J. F. (2005). A novel process for preparation of (1→ 3)-β-D-glucan sulphate by a heterogeneous reaction and its structural elucidation. *Carbohydrate Polymers, 59*(1), 93–99.

Wilhelmson, A., Oksman-Caldentey, K. M., Laitila, A., Suortti, T., Kaukovirta-Norja, A., & Poutanen, K. (2001). Development of a germination process for producing high β-glucan, whole grain food ingredients from oat. *Cereal Chemistry, 78*(6), 715–720.

Wolever, T. M. S., Tosh, S. M., Gibbs, A. L., Brand-Miller, J., Duncan, A. M., Hart, V., Lamarche, B., Thomson, B. A., Duss, R., & Wood, P. J. (2010). Physicochemical properties of oat b-glucan influence its ability to reduce serum LDL cholesterol in humans: A randomized clinical trial. *American Journal of Clinical Nutrition, 92*(4), 723–732.

Wood, P. J., Weisz, J., Fedec, P., & Burrows, V. D. (1989). Large-scale preparation and properties of oat fractions enriched in (1-3) (1-4)-β-D-glucan. *Cereal Chemistry, 66*(2), 97–103.

Yao, N., Jannink, J. L., Alavi, S., & White, P. J. (2006). Physical and sensory characteristics of extruded products made from two oat lines with different β-glucan concentrations. *Cereal Chemistry, 83*(6), 692–699.

Yiu, S. H., Wood, P. J., & Weisz, J. (1987). Effects of cooking on starch and b-glucan of rolled oats. *Cereal Chemistry, 64*(6), 373–379.

Zeković, D. B., Kwiatkowski, S., Vrvić, M. M., Jakovljević, D., & Moran, C. A. (2005). Natural and modified (1→ 3)-β-D-glucans in health promotion and disease alleviation. *Critical Reviews in Biotechnology, 25*(4), 205–230.

Zhang, M. (2010). Heating-induced conformational change of a novel β-(1→ 3)-d-glucan from Pleurotus geestanus. *Biopolymers: Original Research on Biomolecules, 93*(2), 121–131.

# Chapter 11
# Industrial Applications of $\beta$-Glucans

**Reshu Rajput, Balwant Singh, and Mudasir Bashir Mir**

## 11.1 Introduction

$\beta$-glucan, a polysaccharide known for its distinctive molecular structure and diverse biological activities, has emerged as a versatile compound with wide-ranging applications across various industrial sectors. Derived from sources such as oats, barley, mushrooms, and yeast, $\beta$-glucan has gained significant attention for its potential to transform the landscape of foods, medicines, cosmetics, feeds, and health products (Zhu et al., 2015). This comprehensive review explores the intricate web of scientific advancements and industrial breakthroughs that have propelled $\beta$-glucan into the forefront of innovation. The molecular complexity of $\beta$-glucan, characterized by unique $\beta$-glycosidic linkages, sets it apart from other polysaccharides and underlies its multifaceted properties (Jayachandran et al., 2018). Ongoing research continues to unravel the details of its structure, leading to refined extraction methods from different sources, enabling its integration into diverse industrial applications. From functioning as a key ingredient in foods to playing a role in drug delivery systems, $\beta$-glucan has proven to be a valuable asset, contributing to improved product quality, enhanced nutritional profiles, and increased therapeutic efficacy. Most of the prebiotic sources contains B-glucans as their vital component (Slavin, 2013).

This exploration takes us through the realms of foods, medicines, cosmetics, feeds, and health products, revealing the profound impact of $\beta$-glucan in each sector. Its ability to fortify foods, modify textures, and its promising applications in disease

R. Rajput (✉)
University Institute of Agriculture Sciences, Chandigarh University, Mohali, Punjab, India

B. Singh
University Institute of Engineering, Chandigarh University, Mohali, Punjab, India

M. Bashir Mir
Department of Food Science & Technology, Govt. Degree College,
Baramulla, Jammu & Kashmir, India

 
M. Bashir Mir, S. A. Mir (eds.), *Beta-Glucan: Sources, Properties and Applications*,
https://doi.org/10.1007/978-3-031-95788-8_11

prevention and anti-aging formulations position *β*-glucan as a beacon of innovation. β-glucan is recognized as a strong stimulant for boosting immunity and works immeasurably against benign and malignant tumors (Natividad & Verdu, 2013). As we navigate the intersections between science and industry, we address challenges faced by researchers and industries, examining regulatory considerations and envisioning future trends that may shape the continued integration of *β*-glucan into daily lives.

In an era where consumer preferences lean towards health-conscious and sustainable choices, *β*-glucan emerges as a natural and potent solution. Exploring its applications aims to illuminate the potential of *β*-glucan to redefine industries and contribute to the global pursuit of healthier, more sustainable living. Earlier studies conducted on naturally derived β-glucan from cereals and microorganisms reveals its skin health promotion effects including improvement in skin by its attributes like antioxidant, antiwrinkle, wound healing and moisturizing properties (Du et al., 2014). The physical characteristics of β-glucan like viscosity, water solubility and gelation have established it a useful material in the area of food science and nutrition research recently. The subsequent sections of this comprehensive review will unravel the nuances of *β*-glucan's impact on foods, medicines, cosmetics, feeds, and health products, offering a holistic perspective on its role in shaping the future of diverse industries.

## 11.2 Medicinal Applications

*β*-glucan's immunomodulatory properties position it as a crucial component in medicinal applications. Its ability to boost the immune system and reduce inflammation has led to its inclusion in pharmaceutical formulations. Moreover, *β*-glucan shows promise in drug delivery systems, improving the bioavailability and targeted delivery of therapeutic agents. Leveraging *β*-glucan's immunomodulatory properties, the pharmaceutical industry explores its potential for preventing infections and supporting immunotherapies. Recognized for its anti-inflammatory effects, *β*-glucan holds value in formulations aimed at managing inflammatory conditions. As a key component in dietary supplements, *β*-glucan provides a natural means to support immune health, manage cholesterol, and promote overall well-being.

## 11.3 Health Products and Dietary Supplements

In health products, *β*-glucan has found its place in dietary supplements. The formulation of health products harnessing *β*-glucan's properties has gained popularity, offering consumers a natural and beneficial supplement to support well-being. Navigating regulatory considerations is crucial to ensure the safety and efficacy of these products.

The significance of $\beta$-glucan in industrial applications lies in its capacity to deliver functional benefits, nutritional enhancement, and health-promoting effects across diverse sectors. As industries continue to explore and leverage the potential of $\beta$-glucan, it is poised to play a pivotal role in shaping the future of products that prioritize both health and sustainability.

## 11.4 β-Glucans as an Essential Functional Food

Barley and oats are the two chief sources of β-glucans. Although barley is considered as the richest source of β-glucans, oats and its resultant products such as oat bran, oatmeal and flour of oats dragged much attention of researchers (Jayachandran et al., 2018). Oats have proved to have positive impacts in serum cholesterol reduction and prevention of heart diseases (Varma et al., 2016). Widely employed in functional foods and dietary supplements, $\beta$-glucan modifies texture and rheological properties, elevating sensory attributes. β-glucans are known for their beneficial health effects and also for their outstanding properties. Due to its thickening, gelation, stabilizing and emulsification properties, they are widely employed in the preparation of soups, sauces, beverages and various other food products (Zhu et al., 2016). The incorporation of β-glucans into the preparation of cakes and bread enhanced their physical properties and also increased the number of bread loafs (Lante et al., 2023). In the pasta products, β-glucans depicted a low glycemic disease making if effective against various metabolic disorders (Lazaridou & Biliaderis, 2007). B-glucan is mainly recognized for its cholesterol lowering effects and the possible mechanism responsible for reduction in cholesterol level through gut microbiota is chiefly attributed to the production of SCFA (propionate). The microbiota present in the gut metabolizes the fibres and produces the SCFA to the host. The increase in the ratio of propionate to the acetic acid which is the chief substrate for cholesterol biosynthesis, resulted in the reduced cholesterol biosynthesis. Consumption of cereal β-glucans have been resulted in the reduction of plasma cholesterol level and regulation of blood glucose concentration including insulin responses (Hallfrisch et al., 2003; Li & Cheung, 2019). B-glucans from mushroom such as lentinan also contribute hypocholesterolemic effect in humans by reducing the lipoproteins level in blood including both HDL and LDL (Giavasis, 2014).

## 11.5 Structural Features of β-Glucan and Its Chemistry

The unique molecular composition of $\beta$-glucan is characterized by β-glycosidic linkages between glucose molecules. This results in a linear or branched structure, influencing its functional properties and garnering attention from both researchers and industries. Chemically, β-glucans are classified as heterogeneous non-starch polysaccharide or complex carbohydrate, comprises glucose molecules linked by β-glycosidic bonds. Despite of simple monosaccharide composition, β-glucans possess remarkable variability in structure. Depending upon their variable sources, they differ chiefly in degree of branching, types of linkages and molecular weight (Li & Cheung, 2019). Its molecular composition varies depending on the source, found in diverse natural materials like grains, fungi, and bacteria. β-glucans can display a linear or branched configuration, and the arrangement of glucose units contributes to their distinct properties (Du & Xu, 2014). The β-glycosidic linkages, differentiating *β*-glucans from other polysaccharides, play a key role. The specific β-glycosidic linkages involve the anomeric carbon of one glucose unit attaching to the fourth carbon's hydroxyl group in the adjacent glucose unit. This results in β-1,3 or β-1,4 linkage patterns, influencing the overall structure. *β*-glucans exhibit various linkage patterns, including β-1,3, β-1,4, and β-1,6, contributing to structural diversity and functional variations (Table 11.1). Different sources yield distinct types of *β*-glucans. Cereal grains like oats and barley typically contain β-1,3 and β-1,4 linkages, while mushrooms may have β-1,3 and β-1,6 linkages. B-glucan derived from baker's yeast comprises of β-1,3 and β-1,6 linkages (Li & Cheung, 2019), and seaweed may exhibit various linkage patterns.

β-glucans are generally water-soluble, with solubility influenced by factors such as source, molecular weight, and linkage patterns. Degree of solubility and type of linkage is crucial in determining functionality, especially in applications like foods and health products and play a vital role in the immune stimulation by β-glucans. Soltanian et al. (2009) concluded that, inflammatory cytokine production can be strongly induced by soluble glucans as compared to least soluble ones. β-glucans can vary in molecular weight, influencing their physiological effects. Higher molecular weight *β*-glucans are often associated with enhanced immunomodulatory properties, making them particularly valuable in health-related applications (vannucci et al., 2013).

## 11.6 Fortification and Nutritional Enhancement

Fortification and nutritional enhancement involve the deliberate addition of specific nutrients or bioactive compounds to foods or products to elevate their nutritional value and confer health benefits. *B*-glucan, recognized as a soluble dietary fiber with diverse health advantages, offers opportunities for fortification and nutritional enhancement across various product categories. Here's an overview of how *β*-glucan can contribute to fortification and nutritional enhancement. Incorporating oats and barley, rich sources of *β*-glucan, into breakfast cereals, granola bars, and oat-based snacks fortifies these products with heart-healthy benefits and enhances digestive health. Inclusion of *β*-glucan in snack bars or trail mixes amplifies their fiber content, enhancing satiety and contributing to heart health. These products can be positioned as healthier snack options. *β*-glucan is available in supplement form, either consumed directly or added to beverages and smoothies. This offers a convenient way for individuals to increase their fiber intake. Addition of *β*-glucan to baby

**Table 11.1** Sources of β-glucan and their type of linkage

| β-glucan | Source | Type of linkage | Reference |
|---|---|---|---|
| Glomerellan | *Glomerella cingulate* | Branched 1 → 3 & 1 → 6 | Gomaa et al. (1992) |
| Lentinan | *Lentinula edode* | β −1 → 3 with β-1 → 6 | Zhang et al. (2011) |
| Schizophyllan | *Schizophyllum commune* | β −1 → 3 with β-1 → 6 | Tabata et al. (1981) |
| Yestimun | *Saccharomyces cerevisiae* | Branched 1 → 3, 1 → 6 β-glucan | Stier et al. (2014) |
| Pullulan | *Aureobasidium pullulan* | α- (1 → 4), (1 → 6) glucan | Xu and Zhang (2016) |
| Pachyman | *Poria cocos* | β-1 → 3-glucan rarely with few internal (1 → 6) glucan) | Xu and Zhang (2016) |
| Pustulan | *Lasallia pustulata* | β-1 → 3-glucan | Pereyra et al. (2003) and Synytsya and Novak (2014) |
| Krestin PSK | *Trametes versicolor* | β-1 → 4 main cahin with β-1 → 3 and β-1 → 6 side chain | Kurashige et al. (1997) |
| Epiglucan | *Epicoccum nigrum* | 1 → 3 and 1 → 6, β-glucan | Schmid et al. (2001) |
| Grifolan | *Grifola frondosa* | β 1 → 3 glucan | Münz et al. (2005) |
| Scleroglucan | *Sclerotium rolfsii* | β 1 → 3 D-glucose backbone with one β(1 → 6) d-glucose side chain | Survase et al. (2006) |
| Betafectin | *Saccharomyces cerevisiae* | β(1 → 6) branched β (1 → 3) glucan | |
| Zymosan | *Saccharomyces cerevisiae* | Glucose units connected by β-1 → 3 glycosidic linkages | Liu et al. (2011) |
| Curdlan | *Alcaligenes faecalis* | Linear 1 → 3 β-glucan | West (2009) |
| Cereal β-glucan | | | |
| Barley β-glucan | *Hordeum vulgare* | Linear (1 → 3;1 → 4) β-glucan | Papageorgiou et al. (2005) |
| Oat β-glucan | *Avena sativa* | Linear (1 → 3;1 → 4) β-glucan | Charlton et al. (2012) |
| Wheat β-glucan | *Triticum vulgare* | Linear (1 → 3;1 → 4) β-glucan | Li et al. (2006) |

cereals and snacks supports the development of a healthy digestive system in infants, contributing to meeting their dietary fiber needs. Fortifying meal replacement shakes with β-glucan augments their fiber content, promoting a sense of fullness and supporting weight management goals. β-glucan incorporation into specialized health foods, targeting heart health, gut health, or immune support, aligns with consumer preferences for foods offering added health benefits. β-glucan's association with lowering cholesterol levels contributes to cardiovascular health. The soluble fiber in β-glucan supports healthy digestion and may alleviate constipation. β-glucan's potential to regulate blood sugar levels is beneficial for individuals with diabetes or those at risk.

## 11.7 Incorporating β-Glucan in Functional Foods

Integrating β-glucan into functional foods requires a strategic approach to ensure that the final products not only deliver the desired health benefits associated with β-glucan but also align with consumer preferences for taste, texture, and overall sensory appeal. Consider the following considerations and strategies when incorporating β-glucan into functional foods. Choose β-glucan-rich sources such as oats, barley, mushrooms, or yeast based on the targeted health benefits and desired sensory characteristics in the final product. Consider the specific structural and functional properties of the chosen β-glucan source. Variances in solubilities, viscosities, and gelling abilities can influence their performance in different food applications. Determine the optimal dosage of β-glucan to achieve desired health benefits without compromising taste, texture, or overall product quality. Adhere to regulatory guidelines and recommended daily intake levels. Recognize that β-glucan can impact the texture and mouthfeel of foods. Adjustments may be necessary to optimize texture, especially in applications where structure is critical, such as baked goods. Enhance β-glucan solubility through processing techniques or formulation strategies, particularly in applications where a soluble fiber is desired, such as beverages. Although β-glucans generally have a neutral taste, ensure their incorporation complements the overall flavor profile of the product. Consider the specific source of β-glucan and its potential impact on taste. The introduction of β-glucan into food products can exert an influence on their texture and sensory characteristics. As a soluble fiber, β-glucan possesses unique properties that can impact the mouthfeel, structure, and overall sensory perception of food. Here are the primary ways in which β-glucan may shape texture and sensory attributes in food products. β-glucan exhibits gelling properties, making it beneficial in products such as puddings, gels, or thickened beverages where a gel-like texture is desired. Known for increasing viscosity, β-glucan can enhance the thickness and mouth-coating properties of liquids, improving the texture of soups, sauces, or beverages. In some instances, the addition of β-glucan may result in a gum-like texture, influencing the chewiness or stickiness of products like gummies, fruit snacks, or energy bars. Improved Moisture Retention: *B*-glucan's water-holding capacity contributes to better moisture retention in baked goods, resulting in products with a softer texture, increased shelf life, and resistance to staling.

## 11.8 Applications of β-Glucans

### *11.8.1 Food Industry*

In the food industry, β-glucan emerges as a valuable functional ingredient. Its soluble nature facilitates texture and rheological property modifications, making it suitable for incorporation into various food products. Additionally, its health-promoting effects, such as cholesterol reduction and blood sugar management, enhance its role in fortifying foods with added nutritional benefits. β-glucan has several functional properties, including gelation, thickening, stabilizing, and emulsifying (Table 11.2). Scientists have added β-glucan to several food items, such as milk, soups, sauces, and drinks, due to its technological properties. Several studies sought to fully use β-glucan as a bioactive component. For example, because to its health benefits, β-glucan was introduced as tablets or capsules for administration. Integration of $\beta$-glucan into functional beverages like smoothies or fruit juices elevates their fiber content, supporting digestive health and providing a nutritional component. $\beta$-glucan's impact on beverage viscosity influences mouthfeel, especially in products like smoothies, shakes, or fruit juices where a thicker texture is desired. Depending on its source and processing, $\beta$-glucan incorporation into beverages may affect clarity and particle suspension, crucial for maintaining visual appeal. $\beta$-glucan contributes to a creamy texture in oat-based products like oatmeal or cereal bars,

**Table 11.2** Potential applications of β-glucans from different sources in various sectors

| Food | | | |
|---|---|---|---|
| Sources | β glucan name | Applications | References |
| Yeast | Yeast glucan | Salad dressing<br>Cake fillings<br>Frozen deserts<br>Sauce preparation<br>Yoghurt<br>Biscuits and cookies | Thammakiti et al. (2004) |
| Barley and oats | β glucan | Sauce preparation<br>Cake baking<br>Ice creams<br>Yogurts containing β-glucan and pectin | Zhu et al. (2016), Jayachandran et al. (2018), and Rinaldi et al. (2015) |
| Soluble dietary fibers | β –(1,3)(1,4) | Scavenging activity | Kofuji et al. (2012) |
| Eye drops | Mushroom β glucan | Moisture retention | Zhu et al. (2016) |
| Antiseptic creams | Glucan chitosan | Therapeutic wound dressing | Abdel-Mohsen et al. (2016) |
| Cereals | β glucan | Bread<br>Pasta | Zhu et al. (2016), Jayachandran et al. (2018) |
| Lentils | Lentinan | LNT feeding<br>Microbiota exhibit distinctly different space distribution | Jayachandran et al. (2018) |
| Protective creams | β glucan | Therapy of atopic dermatitis | Jesenak et al. (2016) |
| Powders | β glucan | Protective suspensions | Kournikakis et al. (2003) |
| Dietary fibers | β glucan | Feed efficiency<br>Alternative to antibiotic growth promoters | Moon et al. (2016) |

enhancing the overall consistency. Yogurts and Dairy Products: Fortifying yogurt and other dairy products with *β*-glucan enhances their nutritional profile, offering additional fiber and appealing to consumers seeking digestive health benefits. When added to cheese, yoghurt, salad dressings, and ice creams, β-glucan improves rheological properties and reduces calorie content (Guleria et al., 2015). Although β-Glucan is widely investigated in the food industry, little research has been conducted to identify its molecular weight profile in food. Arabic, pectin, and xanthan gum were utilized as beverage thickeners during the close of the twentieth century, with positive results. β-Glucan is a promising bioactive component that can replace fat and thicken food. Numerous research has employed β-glucans in varying concentrations. Various amounts of β-glucans were employed, including 7%, 35%, 50%, and 70% for oats and 50% and 86% for barleys. Other studies isolated β-Glucans from diverse sources or added unspecified amounts of oat flakes to their novel products (Lim, 2018; Kilic & Kankaya, 2016; Vasiljevic et al., 2007; Tudorica et al., 2004).

## 11.8.2 Meat and Dairy Products

Nowadays, the most frequent dairy products seen in supermarkets are yoghurt and cheese. Adding beneficial components like β-Glucan to dairy products is becoming increasingly popular among producers and consumers. Oat dietary fiber, also known as β-Glucan, is a key ingredient in yoghurts and dairy products. Dietary fiber can help speed up the fermentation of yoghurt by increasing the synthesis of acetic and propionic acids. It also improves the yoghurt's texture, consistency, and flavor (Tudorica et al., 2004). It was also employed in an innovative way to make low-fat yogurt and ice cream. Additionally, the potential use of β-D-glucan grains in cheese production was demonstrated. When mixed with soluble dietary fiber, it can be added to low-fat dairy products and cheese curds to increase gelling and improve rheological properties. β-Glucan was employed in both ordinary dairy products and probiotic yogurt (Tudorica et al., 2004). It was previously believed that some beneficial microorganisms in the small intestine were antagonistic to β-glucan. While β-D-Glucan has great gelling capabilities and can mimic fat, it is not recommended for use in meat products like beef burgers or sausages. Adding β-D-glucan to low-fat sausages improves their quality, according to previous studies. Meat Alternatives: Incorporating *β*-glucan into plant-based meat alternatives or utilizing it as a binder in processed meats provides a fiber boost, rendering these products more nutritionally complete. Other researchers have found that adding oat β-D-glucan to lowfat beef considerably improves its quality (Aboushanab et al., 2018).

The chocolate milk is usually produced with the use of carrageenan which acts as a stabilizer & has the ability to reduce the gravitational settling of cocoa powder particles. Chatterjee and Patel (2016) concluded that the 3% incorporation of oat β-glucan along with carrageenan enhances the viscosity, has a positive impact on sensory analysis of chocolate milk and also enhances its dietary fiber content which makes it as functional drink. Vasquez-Orejarena et al. (2018), investigated the effect of oat flour and milk protein on the functional properties of a high protein dairy drink containing atleast 0.75 g of oat β-glucan. They incorporated different combinations of oat flour ranging from 1.50% to 2.30% and milk protein isolate from 2.50% to 4.00%. The final product reflected good suspension stability which was more than 80%. Enhancement in the oat flour significantly enhanced viscosity of the drinks. The incorporation of combination of a protein component together with β-glucan has been evaluated by Temelli et al. (2004) for a formulation of a milk based functional drink. Orange juice-based drink samples consist of 0.5% barley β-glucan and whey protein isolate ranging from 0% to 1.5%. The formulated drink proved to have become more structured, contained less acidity and lost its orange taste with the incorporation of protein and β-glucan content. The possible reason for these alterations is somewhat explained by the ability of β-glucan to neutralize the sour taste, which has been also established by the outcomes of the sensory analysis of β-glucan incorporated pasta (Jaworska et al., 2020), but the mechanism of this needs extensive research.

### 11.8.3 Fermented Dairy Products

Incorporation of oat β-glucan in fermented milk beverages such as acidophillus milk, yogurt, kefir etc. has established it a potential ingredient as it aids in improving physicochemical attributes of the product particularly, enhanced viscosity, reduced acidity and also eliminates consistency problems (Qu et al., 2021). The study conducted by Qu et al. (2021) established that alteration in yogurt formulation with addition of 0.3% level of oat β-glucan partly alters the chemical bonding of three-dimensional mesh structure of yogurt which is attributed to the fact that oat β-glucan retards the interaction with casein, which limits the fermentation process and reduces its duration by about 16 min and enhances mouthfeel and taste of the product. Simultaneously, the study conducted by Jaworska et al. (2020) concluded that addition of oat β-glucan at 1.4% level failed to provide proper structure to yogurt and the product had liquid consistency which can be enlightened by a specific combination of lactic acid cultures, low temperature of fermentation i.e. 36 °C and increased level of polysaccharide which was added to increase the dietary fibre content of the yogurt and to make it a functional product. Poorva Sharma et al. (2017) established that when the mixture containing 70% whey protein concentrate and oat β-glucan was allowed for fermentation, the number of bacterial cells of *Lactobacillus bulgaricus* and *Lactobacillus acidophillus* increased significantly during the initial 10 h of the process. Similar results were confirmed by Angelov et al. (2006) regarding the remarkable influence of oat β-glucan on the activity of

*Lactobacillus plantarum* B28 in the formulation of a new probiotic drink prepared from oats.

### 11.8.4 Bakery Products

Breads and Pastries: Adding *β*-glucan to bread and pastries boosts their fiber content, promoting improved digestive health and aligning with the increasing demand for fiber-rich bakery products. Their appeal as functional foods has spread to bread products, contributing significantly to the surge in interest during the previous two decades. β-Glucan was added to rye flour, rice flour, potato bread, oat-flour bread, and biscuits for testing. Many bread items with the highest bread quality, color, and softness were made after cereal fibers were added to varied degrees (El Khoury et al., 2012). In baked goods such as bread and muffins, *β*-glucan may impart a softening effect, leading to a tender crumb texture. The inclusion of *β*-glucan can influence the handling properties of dough, impacting factors like dough elasticity and extensibility, necessitating formulation adjustments. Incorporating dietary fibers and β-Glucan fortification improved bread's taste, flavor, and structural properties. Oat flour was added to fermented bread in amounts ranging from 2% to 10%. The 10% concentration of β-Glucan in the bread resulted in the best color, flavor, and texture. Others found that adding β-Glucan to biscuits improved taste, adhesiveness, and moisture content. Inglett et al. (2005) demonstrated that Nutrim5®, a β-Glucan product, can be used to manufacture a satisfying pasta with rice and wheat flour. The study found that adding 10% β-Glucan to pasta had no negative impact on the final product (Inglett et al., 2005).

### 11.8.5 Cosmetics and Medicines

B-glucan has documented biological activities which are suitable for skin applications including wound healing, antioxidant activities and anti-wrinkle (Du et al., 2014; Sousa et al., 2023). Recently, some researchers have put β-glucan into the light as an outstanding compound with potential skin benefits, due to its indirect effect on the modulation and stimulation of the immune system and wound healing (Divya et al., 2020). β-glucan's moisturizing attributes contribute to improved skin hydration and barrier function, making it a staple in cosmetics for promoting skin health. Cosmetic formulations capitalize on *β*-glucan's anti-aging properties, addressing wrinkles, fine lines, and signs of aging. In the cosmetic industry, *β*-glucan is recognized for its positive impact on skin health. Its moisturizing properties and contribution to anti-aging formulations make it a sought-after ingredient in skincare products. Native β-glucan has been observed to possess bioactive attributes naturally, but the modification of β-glucan by addition of some functional groups can remarkably enhance solubility and improve bioactive properties (Divya et al., 2020).

Additionally, β-glucan has potential in wound healing and scar reduction, showcasing its multifaceted applications in cosmetics. Potential applications for β-Glucan include medicine and cosmetics (Zhu et al., 2016). β-glucan has been used to make vaccines, treat skin aging, heal eye ulcers, and perform dermatological activities. β-glucan is utilized in cosmetics for its capacity to promote collagen development, reduce wrinkles, and treat eczema. In 2005, Pillai, Redmond, and Roding studied the potential benefits of β-Glucan for skin aging and wrinkle reduction. Research indicates that β-glucan from mushrooms can effectively maintain moisture and be used as eye drops to treat exophthalmia. β-glucan's broad interior cavity has made it a popular wound dressing material in clinical trials. Scientists attempted to combine glucan with chitosan to create a possibly beneficial new therapeutic wound dressing (Delatte et al., 2001). There have also been speculations that it could be used to treat burn-induced oxidative tissue and wound healing. β-glucans in formulations address concerns like acne, eczema, or rosacea due to their soothing and immune-modulating properties. They can be customized blends with hyaluronic acid, vitamins, or botanical extracts offer a comprehensive skincare approach. In addition to its immune-boosting qualities, β-glucan has been used as a strong radical scavenger and potential bone substitute. Soluble dietary fibers, especially β-(1,3), (1,4) Glucans, can significantly reduce blood sugar, insulin, and cholesterol levels, as well as the risk of arteriosclerosis and heart disease (Zhu et al., 2016; Daou & Zhang, 2012).

### *11.8.6 Feeds and Animal Health*

The agricultural sector has witnessed the integration of β-glucan into animal feeds to improve nutritional content and enhance overall livestock health. Beyond nutrition, β-glucan's immunostimulatory effects reduce the reliance on antibiotics in animal farming, aligning with sustainable agricultural practices. Inclusion of β-glucan in animal feeds enhances nutritional content, fostering growth and overall health in livestock. Dietary administration of β-glucans from yeast to lactating ewes resulted in the increase of milk yield by up to 14%. Additionally, the milk composition improved and reported to contain enhanced percentage of fats and proteins (Ząbek et al., 2013; Zaleska et al., 2015). Supplementation of diet of young calves with yeast β-glucan resulted in a much-improved composition of the intestinal microbiota with reduced counts of pathogenic *E. coli* and increased counts of commensal *Lactobacillus* (Zhou et al., 2009; Krüger & Van der Werf, 2018). Moon et al. (2016) conducted a study to study the effect of dietary β-glucan on the performance of broilers and the quality of broiler breast meat and reported that survival rate and feed conversion rate of broilers improved by giving dietary supplementation of 60 ppm β-glucan. These responses depicted that β-glucan can proved to be a effective alternative to antibiotic growth promoter for enhancing growth performance. Another study conducted by Zhang et al. (2020), indicated that 100 and 200 g/T β-1,3- glucan supplementation considering whether heat stress condition or not, can enhanced growth performance without affecting immunity negatively. Under heat stress conditions, the incorporation of β-glucan with same dose level can improve energy digestibility, reduce cooking loss and lower the count of *E. coli* and the concentration of tumor necrosis factor-α (Zhang et al., 2020).

Acting as a prebiotic, *β*-glucan fosters the growth and activity of beneficial gut microorganisms, supporting a healthy digestive system and improving nutrient absorption in livestock. The prebiotic nature of *β*-glucan may aid in decreasing the occurrence of digestive disorders, including diarrhea, by maintaining a balanced and resilient gut microbiome. *β*-glucan's immunomodulatory effects enhance immune cell activity, helping livestock resist infections and diseases, ultimately contributing to overall health. Strengthening the immune system through *β*-glucan supplementation may make livestock less susceptible to stress-related health issues. *β*-glucan improves nutrient utilization in the digestive tract, leading to increased feed efficiency and optimal growth in livestock. Livestock supplemented with *β*-glucan may exhibit improved feed conversion ratios, providing economic advantages for farmers. Livestock supplemented with *β*-glucan may show enhanced resilience to stress, contributing to improved overall well-being and performance. *β*-glucan's potential to stimulate muscle development and protein synthesis contributes to increased weight gain and growth rates in livestock. Livestock on diets enriched with *β*-glucan may demonstrate optimized growth performance, benefiting body weight and muscle development. *β*-glucan may support joint and cartilage health in livestock, addressing issues relevant to certain production systems. Livestock supplemented with *β*-glucan may experience fewer musculoskeletal issues, contributing to improved overall health and mobility.

## 11.9 Medicinal Applications

### *11.9.1 β-Glucan as an Immune System Booster*

*Β*-glucans have gained recognition for their immunomodulatory properties and their potential to enhance the immune system. These intricate polysaccharides are present in diverse sources such as fungi, yeast, oats, barley, and seaweed. Here is an overview of how *β*-glucans may serve as immune system boosters:

### *11.9.2 Activation of Immune Cells*

**Macrophage Stimulation** *β*-glucans interact with immune cells, especially macrophages, enhancing their activity. Macrophages play a vital role in the immune system by engulfing and digesting pathogens.

**Enhancement of Phagocytosis** *β*-glucans boost the phagocytic activity of macrophages, improving the immune system's ability to eliminate foreign particles.

### 11.9.3 Cytokine Production

**Modulation of Cytokines** $\beta$-glucans influence cytokine production, regulating the balance between pro-inflammatory and anti-inflammatory responses. This modulation contributes to a well-coordinated immune function.

**Balanced Immune Response** $\beta$-glucans promote a balanced immune response by influencing the production of key cytokines, maintaining effective immune surveillance.

### 11.9.4 Enhanced Natural Killer (NK) Cell Activity

**Activation of NK Cells** $\beta$-glucans enhance the activity of Natural Killer (NK) cells, crucial immune cells that identify and eliminate infected or abnormal cells.

### 11.9.5 Antioxidant and Anti-inflammatory Effects

**Reduction of Oxidative Stress** Certain $\beta$-glucans, particularly those from mushrooms, possess antioxidant properties, reducing oxidative stress and protecting immune cells.

**Anti-Inflammatory Effects** $\beta$-glucans exhibit anti-inflammatory effects, helping regulate the immune response and prevent excessive inflammation.

### 11.9.6 Enhanced Immunoglobulin Production

**Stimulation of Immunoglobulins** $\beta$-glucans stimulate the production of immunoglobulins, including IgM and IgA, essential antibodies that recognize and neutralize pathogens.

### 11.9.7 Protection Against Infections

**Antimicrobial Properties** $\beta$-glucans demonstrate antimicrobial properties, potentially providing protection against infections and contributing to overall immune defense.

### 11.9.8 Adjuvant in Vaccination

**Use as Vaccine Adjuvants** $\beta$-glucans are explored as vaccine adjuvants, enhancing the effectiveness of vaccines and improving the immune response, showcasing their role in vaccine-mediated immunity.

### 11.9.9 Dose-Dependent Effects

**Optimal Dosage Considerations** The immunomodulating effects of $\beta$-glucans may be dose-dependent, with optimal dosages varying based on the source and the desired immune response.

## 11.10 Conclusion

β-glucan, a polysaccharide known for its distinctive molecular structure and diverse biological activities, holds a significant position in various industrial applications. Abundantly present in sources such as oats, barley, mushrooms, and yeast, the spectrum of applications of β-glucan spreads across different sectors including foods, medicines, cosmetics, feeds, and health products. More research is required to increase the scope of its applications to other fields.

## References

Abdel-Mohsen, A. M., Jancar, J., Massoud, D., Fohlerova, Z., Elhadidy, H., Spotz, Z., & Hebeish, A. (2016). Novel chitin/chitosan-glucan wound dressing: Isolation, characterization, antibacterial activity and wound healing properties. *International Journal of Pharmaceutics, 510*(1), 86–99.

Aboushanab, S., Vyrova, D., & Selezneva, I. (2018, September). Characterization of low-and non-fat yogurt manufactured with addition of beta-glucanas a dietary supplement. In *AIP conference proceedings* (Vol. 2015, No. 1). AIP Publishing.

Angelov, A., Gotcheva, V., Kuncheva, R., & Hristozova, T. (2006). Development of a new oat-based probiotic drink. *International Journal of Food Microbiology, 112*(1), 75–80.

Charlton, K. E., Tapsell, L. C., Batterham, M. J., O'Shea, J., Thorne, R., Beck, E., & Tosh, S. M. (2012). Effect of 6 weeks' consumption of β-glucan-rich oat products on cholesterol levels in mildly hypercholesterolaemic overweight adults. *British Journal of Nutrition, 107*(7), 1037–1047.

Chatterjee, B., & Patel, T. (2016). Increased sensory quality and consumer acceptability by fortification of chocolate flavored milk with oat beta glucan. *International Journal of Clinical and Biomedical Research, 2*, 25–28.

Daou, C., & Zhang, H. (2012). Oat beta-glucan: Its role in health promotion and prevention of diseases. *Comprehensive Reviews in Food Science and Food Safety, 11*(4), 355–365.

Delatte, S. J., Evans, J., Hebra, A., Adamson, W., Othersen, H. B., & Tagge, E. P. (2001). Effectiveness of beta-glucan collagen for treatment of partial-thickness burns in children. *Journal of Pediatric Surgery, 36*(1), 113–118.

Divya, M., Gopi, N., Iswarya, A., Govindarajan, M., Alharbi, N. S., Kadaikunnan, S., et al. (2020). β-glucan extracted from eukaryotic single-celled microorganism Saccharomyces cerevisiae: Dietary supplementation and enhanced ammonia stress tolerance on Oreochromis mossambicus. *Microbial Pathogenesis, 139*, 103917.

Du, B., & Xu, B. J. (2014). Oxygen radical absorbance capacity (ORAC) and ferric reducing antioxidant power (FRAP) of b-glucans from different sources with various molecular weight. *Bioactive Carbohydrate and Dietary Fibre, 3*, 11–16.

Du, B., Bian, Z. X., & Xu, B. J. (2014). Skin health promotion effects of natural beta-glucan derived from cereals and microorganisms: A review. *Phytotherapy Research, 28*, 159–166.

El Khoury, D., Cuda, C., Luhovyy, B. L., & Anderson, G. H. (2012). Beta glucan: Health benefits in obesity and metabolic syndrome. *Journal of Nutrition and Metabolism, 2012*.

Giavasis, I. (2014). Bioactive fungal polysaccharides as potential functional ingredients in food and nutraceuticals. *Current Opinion in Biotechnology, 26*, 162–173.

Gomaa, K., Kraus, J., Rosskopf, F., Röper, H., & Franz, G. (1992). Antitumour and immunological activity of a β 1→ 3/1→ 6 glucan from Glomerella cingulata. *Journal of Cancer Research and Clinical Oncology, 118*, 136–140.

Guleria, P., Kumari, S., & Dangi, N. (2015). β-glucan: Health benefits and role in food industry-a review. *International Journal of Enhanced Research in Science, Technology & Engineering, 4*(8), 3–7.

Hallfrisch, J., Scholfield, D. J., & Behall, K. M. (2003). Physiological responses of men and women to barley and oat extracts (Nu-trimX). II. Comparison of glucose and insulin responses. *Cereal Chemistry, 80*(1), 80–83.

Inglett, G. E., Peterson, S. C., Carriere, C. J., & Maneepun, S. (2005). Rheological, textural, and sensory properties of Asian noodles containing an oat cereal hydrocolloid. *Food Chemistry, 90*(1–2), 1–8.

Jaworska, D., Królak, M., Przybylski, W., & Jezewska-Zychowicz, M. (2020). Acceptance of fresh pasta with β-glucan addition: Expected versus perceived liking. *Food, 9*(7), 869.

Jayachandran, M., Chen, J., Chung, S. S. M., & Xu, B. (2018). A critical review on the impacts of β-glucans on gut microbiota and human health. *The Journal of Nutritional Biochemistry, 61*, 101–110.

Jesenak, M., Urbancek, S., Majtan, J., Banovcin, P., & Hercogova, J. (2016). β-Glucan-based cream (containing pleuran isolated from Pleurotus ostreatus) in supportive treatment of mild-to-moderate atopic dermatitis. *Journal of Dermatological Treatment, 27*(4), 351–354.

Kilic, G. B., & Kankaya, D. A. (2016). Assessment of technological characteristics of non-fat yoghurt manufactured with prebiotics and probiotic strains. *Journal of Food Science and Technology, 53*, 864–871.

Kofuji, K., Aoki, A., Tsubaki, K., Konishi, M., Isobe, T., & Murata, Y. (2012). Antioxidant activity of β-glucan. *International Scholarly Research Notices, 2012*(1), 125864.

Kournikakis, B., Mandeville, R., Brousseau, P., & Ostroff, G. (2003). Anthrax-protective effects of yeast beta 1, 3 glucans. *MedGenMed, 5*(1), 1.

Krüger, D., & Van Der Werf, M. (2018). *Benefits of application of yeast $\beta$-Glucans in ruminants*. Ohly Application Note: Hamburg, Germany.

Kurashige, S., Akuzawa, Y., & Endo, F. (1997). Effects of Lentinus edodes, Grifola frondosa and Pleurotus ostreatus administration on cancer outbreak, and activities of macrophages and lymphocytes in mice treated with a carcinogen, N-butyl-N-butanolnitrosoamine. *Immunopharmacology and Immunotoxicology, 19*(2), 175–183.

Lante, A., Canazza, E., & Tessari, P. (2023). Beta-glucans of cereals: Functional and technological properties. *Nutrients, 15*(9), 2124.

Lazaridou, A., & Biliaderis, C. G. (2007). Molecular aspects of cereal β-glucan functionality: Physical properties, technological applications and physiological effects. *Journal of Cereal Science, 46*(2), 101–118.
Li, X., & Cheung, P. C. K. (2019). Application of natural β-glucans as biocompatible functional nanomaterials. *Food Science and Human Wellness, 8*(4), 315–319.
Li, W., Cui, S. W., & Wang, Q. (2006). Solution and conformational properties of wheat β-D-glucans studied by light scattering and viscometry. *Biomacromolecules, 7*(2), 446–452.
Lim, E. S. (2018). Preparation and functional properties of probiotic and oat-based synbiotic yogurts fermented with lactic acid bacteria. *Applied Biological Chemistry, 61*, 25–37.
Liu, X. L., Lin, N., Zan, D., Yuan, J. J., & Cai, D. L. (2011). Effect of Zymosan on antioxidant and immune function of S 180 tumor-bearing mice. *Cell Biochemistry and Biophysics, 60*, 225–229.
Moon, S. H., Lee, I., Feng, X., Lee, H. Y., Kim, J., & Ahn, D. U. (2016). Effect of dietary beta-glucan on the performance of broilers and the quality of broiler breast meat. *Asian-Australasian Journal of Animal Sciences, 29*(3), 384.
Münz, C., Steinman, R. M., & Fujii, S. I. (2005). Dendritic cell maturation by innate lymphocytes: Coordinated stimulation of innate and adaptive immunity. *The Journal of Experimental Medicine, 202*(2), 203–207.
Natividad, J. M., & Verdu, E. F. (2013). Modulation of intestinal barrier by intestinal microbiota: Pathological and therapeutic implications. *Pharmacological Research, 69*(1), 42–51.
Papageorgiou, M., Lakhdara, N., Lazaridou, A., Biliaderis, C. G., & Izydorczyk, M. S. (2005). Water extractable (1→ 3, 1→ 4)-β-D-glucans from barley and oats: An intervarietal study on their structural features and rheological behaviour. *Journal of Cereal Science, 42*(2), 213–224.
Pereyra, M. T., Prieto, A., Bernabé, M., & Leal, J. A. (2003). Studies of new polysaccharides from Lasallia pustulata (L.) Hoffm. *The Lichenologist, 35*(2), 177–185.
Pillai, R., Redmond, M., & Röding, J. (2005). Anti-wrinkle therapy: Significant new findings in the non-invasive cosmetic treatment of skin wrinkles with beta-glucan. *International Journal of Cosmetic Science, 27*(5), 292–292.
Poorva Sharma, P. S., Neetika Trivedi, N. T., & Yogesh Gat, Y. G. (2017). Development of functional fermented whey-oat-based product using probiotic bacteria. *3 Biotech, 7*(4), 272.
Qu, X., Nazarenko, Y., Yang, W., Nie, Y., Zhang, Y., & Li, B. (2021). Effect of oat β-glucan on the rheological characteristics and microstructure of set-type yogurt. *Molecules (Basel, Switzerland), 26*(16), 4752.
Rinaldi, L., Rioux, L. E., Britten, M., & Turgeon, S. L. (2015). *In vitro* bioaccessibility of peptides and amino acids from yogurt made with starch, pectin, or b-glucan. *International Dairy Journal, 46*, 39–45.
Schmid, F., Stone, B. A., McDougall, B. M., Bacic, A., Martin, K. L., Brownlee, R. T., et al. (2001). Structure of epiglucan, a highly side-chain/branched (1→ 3; 1→ 6)-β-glucan from the micro fungus Epicoccum nigrum Ehrenb. ex Schlecht. *Carbohydrate Research, 331*(2), 163–171.
Slavin, J. (2013). Fiber and prebiotics: Mechanisms and health benefits. *Nutrients, 5*(4), 1417–1435.
Soltanian, S., Stuyven, E., Cox, E., Sorgeloos, P., & Bossier, P. (2009). Beta-glucans as immunostimulant in vertebrates and invertebrates. *Critical Reviews in Microbiology, 35*(2), 109–138.
Sousa, P., Tavares-Valente, D., Amorim, M., Azevedo-Silva, J., Pintado, M., & Fernandes, J. (2023). β-Glucan extracts as high-value multifunctional ingredients for skin health: A review. *Carbohydrate Polymers, 322, 121329.*
Stier, H., Ebbeskotte, V., & Gruenwald, J. (2014). Immune-modulatory effects of dietary yeast Beta-1, 3/1, 6-D-glucan. *Nutrition Journal, 13*(1), 38.
Survase, S. A., Saudagar, P. S., & Singhal, R. S. (2006). Production of scleroglucan from Sclerotium rolfsii MTCC 2156. *Bioresource Technology, 97*(8), 989–993.
Synytsya, A., & Novak, M. (2014). Structural analysis of glucans. *Annals of translational medicine, 2*(2).

Tabata, K., Ito, W., Kojima, T., Kawabata, S., & Misaki, A. (1981). Ultrasonic degradation of schizophyllan, an antitumor polysaccharide produced by *Schizophyllum commune* Fries. *Carbohydrate Research, 89*(1), 121–135.

Temelli, F., Bansema, C., & Stobbe, K. (2004). Development of an orange-flavored barley β-glucan beverage with added whey protein isolate. *Journal of Food Science, 69*(7), 237–242.

Thammakiti, S., Suphantharika, M., Phaesuwan, T., & Verduyn, C. (2004). Preparation of spent brewer's yeast β-glucans for potential applications in the food industry. *International journal of food science and technology, 39*(1), 21–29.

Tudorica, C. M., Jones, T. E. R., Kuri, V., & Brennan, C. S. (2004). The effects of refined barley β-glucan on the physico-structural properties of low-fat dairy products: Curd yield, microstructure, texture and rheology. *Journal of the Science of Food and Agriculture, 84*(10), 1159–1169.

Vannucci, L., Krizan, J., Sima, P., Stakheev, D., Caja, F., Rajsiglova, L., Horak, V., & Saieh, M. (2013). Immunostimulatory properties and antitumor activities of glucans (review). *International Journal of Oncology, 43*(2), 357–364.

Varma, P., Bhankharia, H., & Bhatia, S. (2016). Oats: A multi-functional grain. *Journal of Clinical and Preventive Cardiology, 5*(1), 9–17.

Vasiljevic, T., Kealy, T., & Mishra, V. K. (2007). Effects of β-glucan addition to a probiotic containing yogurt. *Journal of Food Science, 72*(7), C405–C411.

Vasquez-Orejarena, E., Simons, C. T., Litchfield, J. H., & Alvarez, V. B. (2018). Functional properties of a high protein beverage stabilized with oat-β-glucan. *Journal of Food Science, 83*(5), 1360–1365.

West, T. P. (2009). Elevated curdlan production by a mutant of Agrobacterium sp. ATCC 31749. *Journal of Basic Microbiology, 49*(6), 589–592.

Xu, L., & Zhang, J. (2016). Bacterial glucans: Production, properties, and applications. *Applied Microbiology and Biotechnology, 100*(21), 9023–9036.

Ząbek, K., Milewski, S., Wójcik, R., & Siwicki, A. K. (2013). Effect of β-1, 3/1, 6-D-glucan in diet on productivity and humoral and cellular defense mechanisms in sheep. *Acta Veterinaria Brno, 82*(2), 141–146.

Zaleska, B., Milewski, S., & Ząbek, K. (2015). Impact of Saccharomyces cerevisiae supplementation on reproductive performance, milk yield in ewes and offspring growth. *Archives Animal Breeding, 58*(1), 79–83.

Zhang, Y., Li, S., Wang, X., Zhang, L., & Cheung, P. C. (2011). Advances in lentinan: Isolation, structure, chain conformation and bioactivities. *Food Hydrocolloids, 25*(2), 196–206.

Zhang, S., Ou, J., Luo, Z., & Kim, I. H. (2020). Effect of dietary β-1, 3-glucan supplementation and heat stress on growth performance, nutrient digestibility, meat quality, organ weight, ileum microbiota, and immunity in broilers. *Poultry Science, 99*(10), 4969–4977.

Zhou, Y., Diao, Q., Tu, Y., & Yun, Q. (2009). Effects of yeast β-glucan on gastrointestinal development in early-weaning calves. *Chinese Journal of Animal Nutrition, 21*(6), 846–852.

Zhu, F. M., Du, B., Bian, Z. X., & Xu, B. J. (2015). Beta-glucans from edible and medicinal mushrooms: Characteristics, physicochemical and biological activities. *Journal of Food Composition and Analysis, 41*, 165–173.

Zhu, F., Du, B., & Xu, B. (2016). A critical review on production and industrial applications of beta-glucans. *Food Hydrocolloids, 52*, 275–288.

# Index





M. Bashir Mir, S. A. Mir (eds.), *Beta-Glucan: Sources, Properties and Applications*, https://doi.org/10.1007/978-3-031-95788-8

The manufacturer's authorised representative in the EU is Springer Nature Customer Service Centre GmbH, Europaplatz 3, 69115 Heidelberg, Germany. If you have any concerns regarding our products, please contact ProductSafety@springernature.com

Printed and bound by CPI Group (UK) Ltd, Croydon, CR0 4YY
30/09/2026
02225562-0003